SpringerBriefs in Agriculture

For further volumes:
http://www.springer.com/series/10183

Gideon Ladizinsky

Studies in Oat Evolution

A Man's Life with *Avena*

 Springer

Gideon Ladizinsky
Faculty of Agriculture
Hebrew University
76100 Rehovot
Israel

ISSN 2211-808X   ISSN 2211-8098   (electronic)
ISBN 978-3-642-30546-7 ISBN 978-3-642-30547-4   (eBook)
DOI 10.1007/978-3-642-30547-4
Springer Heidelberg New York Dordrecht London

Library of Congress Control Number: 2012938868

Frontispiece: Avena insularis

Printed on acid-free paper

Springer is part of Springer Science+Business Media (www.springer.com)

To my family

Preface

In 1964 I began a study on evolution and species relationships in the genus *Avena*. I could not imagine then that this journey would last for more than 45 years. The genus *Avena* is economically important as it contains the common oat which is grown for human consumption and animal feed and the slim, or sand oat, which is grown for fodder. The subject of my study, however, was the wild oat species because at that time these were not well understood compared to the other cereals and definitely when compared to the present knowledge. It is interesting to note that in addition to the accumulation of that knowledge, between 1967 and 1996, seven new species of wild oats were discovered—one by US oat workers, another three by Canadians, and three by me. I do not know of any other crop plant in which such botanical and genetic progress has been achieved over such a short period.

For about 20 years, from 1973 to 1994, besides my work on oats, I was intensively engaged in a study of the origin and wild relatives of the Middle East pulses, including broad bean, chick pea, lentil, fenugreek, common vetch, and bitter vetch. The work involved collecting wild relatives from various Geographical areas, a preliminary survey of their seed protein profile, performing intra- and interspecific hybridizations and chromosome pairing analyses of their hybrids, and establishment of methods for gene transfer from certain wild relatives to the domesticated forms. I refuted a hypothesis that the broad bean originated from the wild species *Vicia galilea* or any other species of Section Faba. I established crossability relations between the annual species of the genus *Cicer* and discovered the wild progenitor of chick pea in eastern Turkey. In the course on extensive studies of the genus *Lens*, I explored the areas and collected seeds where its wild relatives have been reported in Mediterranean countries, central Asia, Uganda, Ethiopia, and Morocco. These activities yielded two previously unknown species. In addition, following extensive hybridization experiments on the cytogenetic analyses of hybrids, I was able to establish evolutionary relationships between the various species. The use of embryo culture enabled me to hybridize species which otherwise are cross- incompatible. By employing characteristics that could not be selected or influenced by human, I showed that the wild lentil stock which gave rise to the domesticated lentil occurs in northern Syria and southern Turkey.

This small book resulted from conversations with oat specialists who were trained mainly as molecular biologists. They urged me to summarize my experience with oat and my insights and viewpoints regarding several evolutionary issues in that genus. I hope that future generations of oat scientists will benefit from this publication and will add further evidence to deepen our understanding of oat species origin and evolution.

Chapter 1 of this book deals with the morphology and the taxonomy of the genus *Avena*. The morphological characteristics that are employed for identification of the various species are described and evaluated. Taxonomic treatments of the genus are briefly described and their advantages and drawbacks are pointed out. The species concept is discussed and the biological species type seems preferable over the classical morphological concept. Accordingly, a list of the *Avena* species is presented with a key for their identification and the description of each of them.

Chapter 2 describes my research findings in the genus *Avena* from the initial steps of becoming acquainted with the various species, their morphological peculiarities, geographic distribution, and ecological preferences. Considerable effort has been devoted to study the relationships between diploids and tetraploids in Series Eubarbatae. By combined cytological and morphological evidence I successfully used the bristle length at the tip of the lemmas to separate diploids from tetraploids in living material and herbarium specimens and to determine the geographic distribution and the ecological preferences of each of them. While the origin of the tetraploid forms is believed to be through allopolyploidy, the evidence which I presented indicates an autopolyploid origin. In this chapter the newly described species in Section Eubarbatae are described and evaluated.

Chapter 2 also deals with Section Denticulatae to which the common oat belongs. The process by which the genome designation of this hexaploid oat has been established is described and critically assessed and refuted. In this Section, several new species were described in the last 50 years, one of which appears to be the tetraploid progenitor of the hexaploid oats. The discovery of each of these species is briefly described.

Another section of this chapter deals with oat domestication and in particular the recent domestication of the protein-rich tetraploid wild oat *Avena magna*. The latter is a project I have been conducting for the last 25 years. Its purpose is to transfer the domestication syndrome of the common oat to this wild oat, thereby creating a new protein-rich tetraploid oat cultivar.

The last section deals with collecting wild oat species as genetic resources. While collecting wild oats is mentioned throughout the book; in this chapter the methodology which I have developed over 45 years is described.

I am indebted to Prof. Abbo Shahal and Dr. Mike Leggett for their valuable suggestions on the manuscript.

Rehovot   Gideon Ladizinsky

Contents

Chapter 1
Oat Morphology and Taxonomy

Abstract The morphological characters used for classification and identification are those of the spikelet. Additional characters that may be critical for accurate identification are chromosome number and potential for hybridization with other species. The taxonomy of the genus *Avena* in this book is based on that of Malzew, with a number of important changes such as omission of the taxon *A. macrostachya* from the genus. Determination of the *Avena* species is based on the biological species concept, which reflects mainly evolutionary and genetic features together with morphological and ecological uniqueness. Accordingly, a key to the various species is presented, and this is followed by a description of each species.

Keywords Oat morphology · Taxonomy · Morphological species · Biological species · Gene pool system · Identification key · Description

1.1 Oat Morphology

Oats are annual grasses with flat leaf blades; inflorescences open, effuse, or contracted or one-sided panicles with peduncles of pedicellate spikelets. Peduncles at the lower part of the panicle are usually longer bearing several spikelets. Spikelets are large, one to several flowered, and hermaphrodite. The rachilla is fragile, at least below the lowest floret in the wild species but tough in cultivated oats. Glumes, lower and upper, equal to one another or markedly unequal, chaffy, and remain attached to the panicle after seed dispersal. Lemmas are coriaceous to crustaceous, hairy or naked, seven-nerved, two-lobed, or entire with a stout geniculate awn issuing from the dorsal surface. The callous of the disarticulated florets have vertical or oblique scars. The palea is two-keeled, bearing hairs on the keels. There are three stamens, ovary is villous, and there are two lodicules, ovate to lanceolate, acuminate, fleshy below (Fig. 1.1). Grains are oblong, hairy, adherent to the lemma

G. Ladizinsky, *Studies in Oat Evolution*, SpringerBriefs in Agriculture,
DOI: 10.1007/978-3-642-30547-4_1, © The Author(s) 2012

Fig. 1.1 Schematic diagram of the oat flower, **a** partly dissected oat flower at anthesis, **b** longitudinal, and **c** transverse diagrams of the flower, **d** spikelet

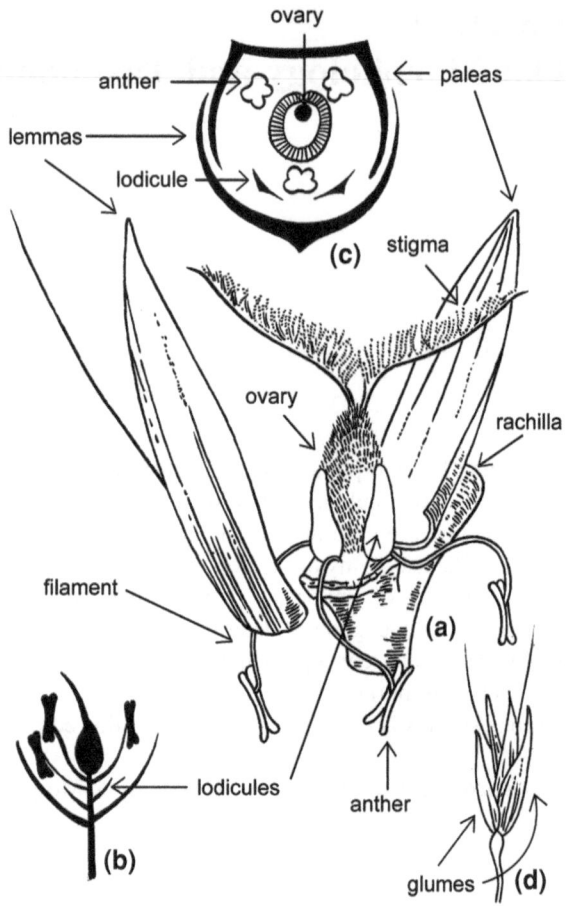

and palea, or free in some cultivated oats. Embryos are about one-eighth the length of the grain with basal hillum.

Of the oat morphological characters, those of the spikelet have been mainly used for species delimitation and classification and can be regarded as diagnostic characters.

They are as follows:

- Glumes shape.
- The structure of the lemma tips.
- Size and shape of the disarticulation scar.
- Shape of the callus at the bottom of the dispersal unit.
- Point of insertion of the awn into the lemma.
- The mode of the spikelet disarticulation at maturity.

In most oat species, the two glumes of the spikelet are equal in length or nearly so. Exceptions are *A. clauda* in which the lower glume is about half as long as of

Fig. 1.2 Variation of glumes shape. **a** Markedly unequal, *A. clauda*, **b** moderately unequal, *A. ventricosa*, **c** equal or nearly so, *A. longiglumis*

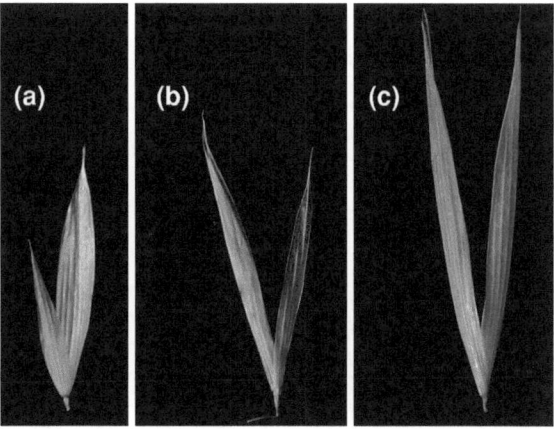

the upper glume, and *A. ventricosa* in which the lower glume is about three-quarters of the length of the upper glume (Fig. 1.2).

Lemma tips have two basic structural forms, the lobes each terminate in a bristle whose length differs. It may have a side tooth, but its appearance is not consistent among plants or even among different spikelets of the same plant, and is thus of little or no taxonomic value. Aristulate lemma tips are typical of *A. clauda*, *A. strigosa*, *A. prostrate*, *A. damascena*, *A, longiglumis*, and *A. barbata*, but in *A. ventricosa* they are subulate (Fig. 1.3). In the other species, the two lobes terminate in denticulate membranous structure. This is seen in *A. canariensise*, *A. agadiriana*, *A. magna*, *A. murphyi*, *A. insularis*, and *A. sativa*.

The size of the disarticulation scar and the callus at the bottom of the dispersal unit is usually interrelated. They are narrow in *A. clauda*, longer in *A. longiglumis*, and the longest in *A. ventricosa*, but in the last species the disarticulation scar, although very long, is remarkably narrow. In *A. strigosa*, *A. barbata* and their allied species the callus is shorter and the disarticulation scar is more elliptical. The largest disarticulation scar is found in *A. magna* and somewhat smaller in *A. murphyi* and the *sterilis* form of *A. sativa*. *A. insularis* can be distinguished from them by its oblong disarticulation scar (Fig. 1.4).

In most oat species the awn is inserted at a point between one-third and one-half of the length of the lemma. *A. murphyi* is unique in that its awn is inserted at a point in the lowest quarter of the lemma, and in *A. ventricosa* the insertion point is in the uppermost quarter of the lemma. (Fig. 1.5).

At maturity wild oat species present one or two modes of spikelet disarticulation; either at the base of the lower floret only, or at each floret (Fig. 1.6). In the first type the dispersal unit contains two or more florets (seeds) and in the second one the dispersal unit is single seeded. This character has been widely used by oat taxonomists because of the ease with which the two modes can be distinguished. The disarticulation mode may be the main or even the only difference between closely related species as in the case *A. clauda* and *A. eriantha* (in which the

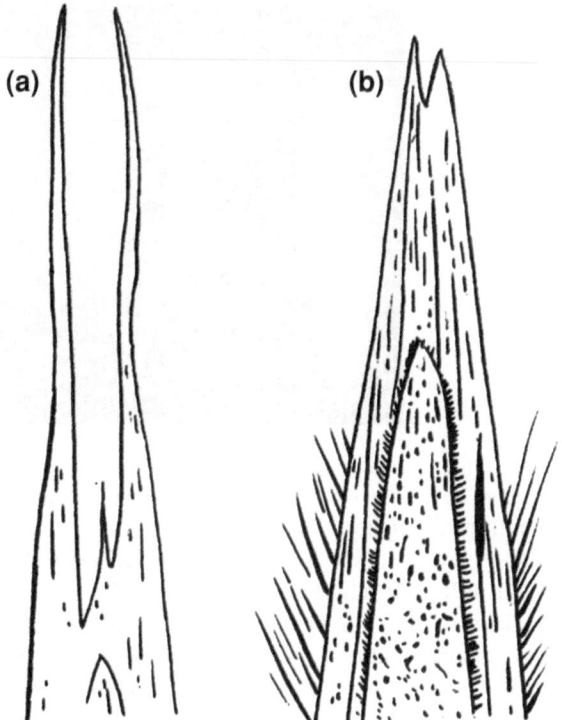

Fig. 1.3 The two basic forms of lemma tips in *Avena*

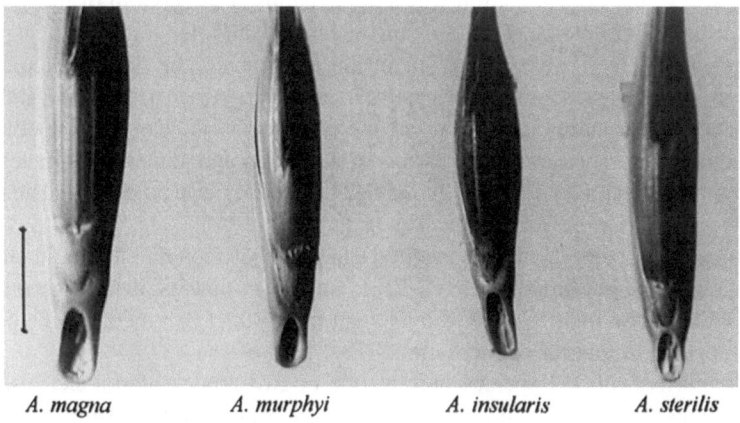

A. magna   *A. murphyi*   *A. insularis*   *A. sterilis*

Fig. 1.4 Disarticulation scar of *A. mgna, A. murphyi, A. insularis*, and *A. sativa* ssp. *sterilis*

dispersal unit is the floret and spikelet, respectively), or of *A. fatua* and *A. sterilis*. In both cases the mode of spikelet disarticulation is governed by a single gene (Rajhathy and Thomas 1967; Coffman 1961) and in each case the two species are

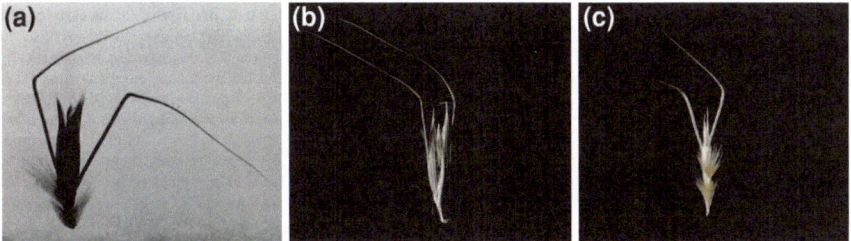

Fig. 1.5 Variation in awn insertion into the lemma, **a** lowest quarter, *A. murphyi*, **b** middle, *A. magna* (hairs removed), **c** upper quarter, *A. ventricosa*

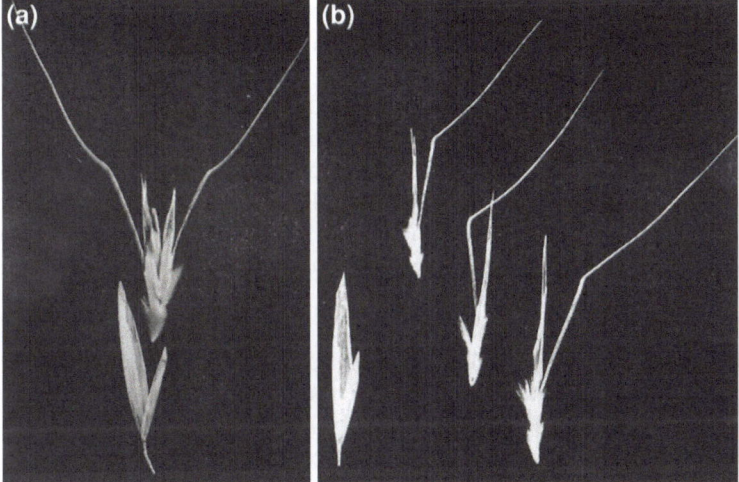

Fig. 1.6 The mode of spikelet disarticulation in *A. clauda*, **a** ssp. *eriantha*, **b** ssp. *clauda*

fully interfertile, casting doubt on the validity of their classification as different species.

In addition to the characters mentioned above, Baum (1977) also employed lodicule and epiblast (supposedly a rudimentary cotyledon) shape in his classification. I tried to examine these microscopic characters but found them difficult to handle and more important, not reliable. I am not aware of anyone other than Baum who uses these traits for oat identification and classification.

A great value for oat classification is the chromosome number. This is $2n = 14$ in the diploid species, $2n = 28$ in the tetraploids, and $2n = 42$ in the hexaploids. The polyploid oats are allopolyploids which are believed to have evolved from interspecific hybridization of species with lower chromosome numbers followed by duplication of the chromosome number of the sterile interspecific hybrid.

In certain cases oat plants may have the same chromosome number and the morphological difference between them is inconclusive. Therefore, the only way to resolve their identity in such cases is to cross them with one another or with tester lines. This was done to distinguish *A. prostrata* and *A. damascena* from each other and from the wild forms of *A. strigosa*.

1.2 Oat Taxonomy

The purpose of plant taxonomy is to arrange the plants of the world in forms of species, genera, families, etc. The basic unit of taxonomy is the species but what constitutes a species is far from clear. Nevertheless, it is commonly accepted that individuals of a certain species represent not only morphological coherence but also evolutionary ties (more about this in Sect. 1.3). In classical taxonomy the basis for species delimitation is according to morphological characters. Appropriate selection of the diagnostic characters for intraspecific, and intrageneric, classification would determine the quality of the classification and the validity of the resulting taxa. Species are usually differentiated from one another by definitive characters (not intermediate) or combination of characters.

A species name is composed of two words, the genus name and the species epithet like, *Avena sativa*. The species name stands for all the peculiarities of a particular species and is of immense value for communication between botanists and scientists.

Within a genus the species may be grouped in sections, subsections, and series. Here too, the grouping is usually according to morphological resemblance. Similarly, intraspecific classification might be based on morphological characters or geographic distribution. The most common taxon here is the subspecies.

During the last 80 years two taxonomic treatments of the genus *Avena* were published by Malzew (1930) and Baum (1977). Malzew was an oat specialist of Vavilov's team who published a series of monographs on several crop plants and their wild relatives. Obviously, Malzew did not have the botanical and cytogenetical knowledge of oats that is available now, but his attitude toward oat classification is interesting.

Malzew arranged the genus *Avena* in two sections: Avenastrum with a single perennial species *A. macrostachya* and Euavena where he placed all the annual oat species. In fact his taxonomical treatment deals only with the Euavena section. In addition to its perennial growth habit, *A. macrostachya* differs from other oat species in being autotetraploid, outcrosser, and self-incompatible, and under normal circumstances it is reproductively isolated from the annual species, although hybridization between them may be achieved by the aid of embryo rescue techniques. To my mind it would be more appropriate to transfer this taxon to the genus *Helicotrichon* as was suggested by a number of botanists in the past. There is no doubt that transfer of *A. macrostachya* to the genus *Helicotrichon* would make the *Avena* genus more uniform and coherent.

Section Euavena was further divided by Malzew into two subsections: Aristulatae which was further divided into three series, Inaequaliglumes, Stipitatae, and Eubarbatae and the other subsection, Denticulate, which was not divided into series.

Except for series Eubarbatae, each of the other series and subsection Denticultae are composed of two species each. Disarticulation of the dispersal unit is at the base of the lower floret in one species and at the base of each floret in the other. While that grouping seems natural in series Inaequaliglumes and subsection Denticulatae (because the two members are interfertile and grow in the same habitats), it is not natural in series Stipitatae. The two species in that series, *A. longiglumis* and *A. ventricosa*, are cross incompatible and occur in different habitats. On the other hand, the former species is cross compatible with members of series Eubarbatae and the latter with members of series Inaequaliglumes. It would thus be appropriate to transfer them to these respective series and to delete series Stipitatae.

While Malzew's attitude toward oat classification is reasonable, some changes should be made in order to make it compatible, as much as possible, with present botanical and genetic knowledge and are as follows:

1. As already mentioned the genus *Avena* will include only the annual species.
2. The two sub sections, Aristulatae and Denticulatae, should become sections.
3. Section Aristulatae should be divided into two Series, Inaequaliglumes and Eubarbatae.

This grouping will be used from here on in this book.

During the 47 years between the publications of the two taxonomic treatments of the genus *Avena* several new oat species were discovered and a considerable amount of genetic and cytogenetic data have been accumulated on species relations in this genus. Therefore, the expectation from Baum's treatment was that a new overall treatment would bring a more logical approach to oat classification and adequate synthesis between morphological evidence and the growing amount of genetic data on species relationships in the genus *Avena*. However, this has not materialized. As T. Rajhathy commented to me after Baum's monograph was published, "this is not the baby I was hoping for".

Baum arranged the genus *Avena* in six sections and 27 species. Unlike Malzew he correctly placed *A. ventricosa* in the same section with *A. clauda* and *A. eriantha*, and *A.longiglumis* in the same Section that includes members of Malzew's series Eubarbatae.

Baum's classification was based almost entirely on morphological characters, and so some new species were described for the first time in his monograph. When he used data of more genetic nature, they often contradicted the morphological evidence. He ignored the fact that *A. clauda* and *A. eriantha* are interfertile as are *A. sativa, A. sterilis*, and *A. fatua* and that the different fruiting morphs are due to single genes (Rajhathy and Thomas 1967, Coffman 1961). The genome labeling of the various species in the monograph is also a good example of the contradiction between the morphological attributes Baum used for describing various species and the genomes he assigned to them.

A genome relates to the chromosome characteristics of the gamete in terms of number, size, shape, and linear arrangement. By definition, two individuals share the same genome if their hybrid develops normally, exhibits normal chromosome pairing at meiosis, is fertile, and no breakdown occurs in the segregating generations. According to Baum *A. clauda* and *A. eriantha* have different genomes. On the other hand some of the Baum's species share the same genome and are interfertile. So why are they are classified as different species?

The confusion in Baum's monograph is even greater among the diploids classified under the A genome, original or modified. The species *A. brevis, A. hispanica,*, and *A. nuda* are interfertile were labeled by Baum as sharing the A genome, but *A. canariensis* which is intersterile with them and with other species of the A genome like *A strigosa* and *A. wiestii* has also been labeled as having A genome.

The tetraploid species *A. barbata, A. abyssinica*, and *A. vaviloviana* are interfertile and were regarded by Baum as separate species, but they all share the same AB genome.

The greatest confusion of species with different names but sharing the same genome is in the hexaploid oats. All the hexaploid oats are interfertile and according to Baum (1977) share the same ACD genome. So among the hexaploid oats he recognized only one cultivated species, *A. sativa* but six wild hexaploid species.

As already mentioned, Baum (1977) employed lodicules and epiblast shape in his classification. I found these microscopic characters difficult to handle and often unreliable. More important, from Baum's data (1977) it appears that among the cultivated diploid species *A. hispanica, A. brevis, A. strigosa*, and *A. nuda*, which are all apparently interfertile, there are two lodicule types; fatua-type for the former species and strigosa-type for the latter two which suggests intraspecific variation.

The same is for the tetraploid species (*sensu* Baum) *A. barbata, A. abyssinica*, and *A. vaviloviana*. These three oat types are interfertile but the former and the latter ones have fatua lodicule-type while the middle one a sativa-type.

All the hexaploid oat species listed by Baum are interfertile, yet their lodicules and epiblast are of several types, indicating, again, intraspecific variation.

It seems, therefore, that variation in lodicules and epiblast shape among taxa which are interfertile indicates that they cannot be used as diagnostic characters.

1.3 The Species Concept in *Avena*

My experience with the taxonomy of *Avena* and the genetic properties of the various species led me to reconsider the species concept in general and particularly in *Avena*. Specifically in genera of economic importance, classical taxonomy is redundant if it does not reflect the evolutionary and genetic ties among the various taxa. The reason is that the wild members of such genera are potential source of genetic diversity for the improvement of the domesticated species and the genetic relations between them is an indication for the possibility of their exploitation for that purpose. Therefore, delimiting of species boundaries is not taxonomic game but represents a reasonable effort to combine morphological and genetic characteristics which express evolutionary ties. Such approach would satisfy both botanists and plant geneticists and breeders.

The species is a basic unit in plant and animal classification and a corner stone in biology and evolution. However, strange though it may seem, there is no single and common definition of the term species.

The naïve approach to the term species is that it is something real that scientists merely described. The truth is that this term is an invention by scientists. It is a subjective concept and some biologists may deny the validity of certain species that are accepted by others. In addition, there are several types of species. The most common are the morphological species and the biological species.

1.3.1   The Morphological Species

The morphological species is determined almost exclusively by morphological characters. Accordingly, a species is a group of individuals that share the same key characters known as diagnostic characters. The diagnostic characters which are used to separate species within a genus are of discontinuous nature with no intermediate state. However, some variations do exist in many such characters. Differentiation between closely related species in terms of a single character or combination of characters can be done, or at least has been attempted in the past.

Historically and currently, this is the first kind of species that was employed for the purpose of taxonomy. Most of the known species are in fact morphological species, and in many cases their legitimacy stands even when more rigorous measures of identification and classification, such hybridization experiments, are applied. The main advantage of utilizing the morphology for species identification is that no experimental procedure is involved or required.

The main disadvantage of the morphological species is that the genetic properties are hardly used or taken into consideration. The discontinuous nature of a particular trait, though an important diagnostic character, might indicate that it is governed by a single gene. This is the case with *A. clauda* and *A. eriantha* as already mentioned. These two are distinguished from one another by their dispersal units, the floret in the former and spikelet in the latter. However, they grow in mixed stands in most of their distribution area, are interfertile, and the difference in the mode of seed dispersal is governed by a single gene. It follows that these two are legitimate species by their morphology, but not by their evolutionary ties.

Another drawback of the morphological species becomes apparent in what is called "sibling species". These are species with the same, or nearly the same morphology but are isolated from one another by various mechanisms such as cross-incompatibility, hybrid sterility, or hybrid breakdown in segregating generations. In series Eubarbatae, the diploid forms, *A. damascena* and *A. prostrata* form a group of sibling species. Morphologically it is difficult to distinguish between them but they are reproductively isolated from each other. Therefore, the safest way to distinguish between them is to cross the two.

1.3.2 The Biological Species

The biological species has been proposed in order to give more meaning to the genetic properties of the species concept. Accordingly, a biological species is a group of individuals which actually or potentially interbreed and form common gene pool which is isolated from other gene pools by means of reproductive barriers (not geographic barriers). This definition also provides the testing tool for determining if two individuals are members of the same species or are of different species, namely, hybridization. The two extreme cases in such an experiment are that the hybrids between the two individuals are either fertile or sterile. Obviously, when hybrids between two morphological species are sterile they automatically can be regarded as members of different biological species. However, if the hybrids are fully fertile, the two morphological species can be regarded as members of the same biological species. The difficulty is when these hybrids are only partially fertile. In those cases more criteria must be taken into account in order to determine their taxonomic status. A major consideration is to what extent the partial fertility is in the range accepted for intraspecific variation or exceeds it.

Chromosome variation is common in *A. barbata*. It can be detected in hybrids among populations or even among plants of the same population that exhibit multivalent association in meiosis. These multivalent are the result of reciprocal translocations between non-homologous chromosomes in the parents. Usually such multivalent association causes reduction of fertility. Despite that, these plants are regarded as members of the same species. Hybrids between *A. barbata* and *A. abyssinica* or *A. vaviloviana* also show 1–2 multivalent associations at meiosis and their fertility is only slightly reduced. Thus, these forms cannot be regarded as separate species even though they are morphologically distinct from *A. barbata* by a character that evolved under human influence and is governed by four genes.

The biological species concept has been widely used in crop plants and their wild relatives. This becomes possible because many hybridization experiments carried out by plant geneticists and plant breeders revealed genetic affinities between morphological species and clearly indicated the taxa from which gene transfer to the cultigen could be made. In many crop plants this has entailed drastic reduction in the number of species. In wheat, for example, Jakubziner (1958) recognized ten cultivated tetraploid and seven hexaploid species while Morris and Sears (1967) reduced the numbers to two tetraploids and a single hexaploid species.

Although interfertility is the most critical characteristic of a biological species, morphology and ecological preferences are also relevant properties of this entity.

After all, most of the biological species are recognized by their morphological characters but alone are not always sufficient to delimit species boundaries. Thus, for example, the diploid forms of series Eubarbatae, wild and cultivated, are all members of the same species. They include the taxa *A. brevis, A. hispanica, A. nuda, A. strigosa* (cultivate forms) and the wild *A. hirtula,* and *A. wiestii.* They are all interfertile and separating them to different species is like considering black-eyed and blue-eyed individuals as members of separate species. Why should something that is an obvious absurdity for mankind is legitimate for plants? I regard all these diploid forms collectively as members of *A. strigosa.*

To this group of taxa one should also add the taxon *A. atlantica* from southern Morocco which was described about 25 years ago. It grows mainly on sandy soil and its main characteristic is the dispersal unit which comprises the entire spikelets (glumes excluded). It was collected independently by G. Fedak, M. Leggett, and P. Hagberg, but was described by the former together with B.R. Baum.

I crossed *A. atlantica* with members of *A. strigosa.* The hybrids were normal and fertile and segregation regarding the kind of the dispersal unit in F2 indicated the involvement of two genes.

The rare occurrence of forms with morphological characteristic which is not typical of series Eubarbatae, but are found in other oat species, in *A. clauda, A. magna, A. murphyi, A. insularis* and in the hexaploid oats, is a reminder of Vavilov's law of homologous series. According to this law a character which is common in one species may spontaneously occur in a related species or even in a closely related genus. Another example of this law in oats is the seed-retaining property of cultivated oats, diploids, and hexaploids. Furthermore, the occurrence of the two modes of seed dispersal in the taxa *A. clauda,* and *A. eriantha* and among members of *A. sativa* indicate the polyphyletic origin of the spikelet dis-articulation mode in the genus *Avena.*

1.3.3 The Gene Pool System

Plant breeders and plant geneticists are usually interested in crop wild relatives as sources of genetic variability which does not exist in the crop plants. For the sake of exploring this kind of diversity and transferring it to cultivated background they must be familiar with genetic affinity between the particular wild species and the cultigen. This information is not automatically provided by the morphological species concept. On the contrary, the existence of a large number of morphological species may discourage plant breeders from seeking genetic diversity in them because they are not sure about the feasibility of gene transfer from them to the cultivated material. To overcome this difficulty Harlan and de Wet (1971) proposed the gene pool system in which crop plants and their relatives, cultivated and wild, are arranged in three gene pools (GPs).

The first is GP1, in which all the members are interfertile. These members include the various forms of the crop plant, regardless of their taxonomic status and also its wild progenitor which may also be classified by taxonomists to several species.

GP2 includes taxa from which some gene can be transferred to the cultigens despite some hybrid lethality or sterility. Even when some seeds are set in the interspecific hybrid, transferral of a specific character is not automatically guaranteed because it might be located on non-homologous chromosome segments.

GP3 includes all other related taxa from which gene transfer to the cultigens is not possible. However, these taxa are the target of exploring and employing new techniques that would eventually enable gene transfer.

All the members of *A. strigosa*, wild and cultivated, are included in the cultigen's GP1 of this species. Similarly, all the hexaploid oats are members of the common oat *A. sativa* GP1.

GP2 of *A. strigosa* includes *A. barbata* and its related tetraploid forms. Gene transfer from them can be made through a triploid bridge. Using such a bridge, I transferred to var. Saia of *A. strigosa* resistance to the herbicide Atrazine from *A. barbata*. This resistance was detected in Israel by B. Rubin and his associates along a road that had been routinely sprayed by atrazine for many years. It is a maternally inherited character and I used *A. barbata* as the female parent. The triploid hybrids produced only a few seeds and from these I selected a diploid plant which was similar to var. Saia and also atrazine resistant.

All other oat species are in GP3 of *A. strigosa* because of various obstacles for hybridization.

Gp2 of the common oat includes mainly the tetraploid species *A. magna*, *A. murphyi* , and *A. insularis*.

It is my belief and conviction that utilization of the biological species concept for oat classification is the most appropriate and reasonable approach for botanists as well as geneticists and plant breeders. Below is the list of the biological species of the genus *Avena*. Names of some of the morphological species have been omitted or retained as subspecies.

1.4 The Biological Species of *Avena* and Their Main Sub Species

For reasons already indicated above, it seems to me that adopting the biological species concept for oat classification is much more beneficial to botanists and plant breeders alike. Below is the list of the *Avena* species known at present, and their main sub-pecies (Table 1.1).

While lowering the taxonomic status of a number of *Avena* taxa from species to subspecies is logical, they are occasionally written as species in this book. Particularly in events described prior to this publication.

Table 1.1 The biological species of *Avena* and their main sub species

Species name	Sub species	Chromosome no.
A. clauda	clauda	14
	eriantha	
A. ventricosa		14
A. longiglumis		14
A. prostrata		14
A. damascena		14
A. strigosa	strigosa	14
	wiestii	
	hirtula	
	atlantica	
A. barbata	barbata	28
	abyssinica	
	vaviloviana	
A. canariensis		14
A. agadiriana		28
A. insularis		28
A. murphyi		28
A. magna	magna	28
A. sativa	sativa	42
	sterilis	
	fatua	

1.4.1 Key to the Avena Species

The key is based on several morphological characters, chromosome number, ecological preferences, and the place of origin of the oat material. The morphological traits used in the key are shape of the glumes, shape of the spikelet, structure of the callus at the lower part of the dispersal unit, shape of the disarticulation scar, point of awn insertion into the lemma, and structure of the lemma tip. Differentiation to subspecies is made mainly according to the mode of spikelet disarticulation.

Note: (1) Counting of chromosome numbers may be necessary for separating *A. barbata, A. prostrate* and *A. damascene* from one another. (2) Confirming the identity of the latter two species may require hybridization with known accessions of these species.

1.4.1.1 Key:

1. Glumes markedly unequal in size (2)
_ Glumes equal or almost equal in size (4).

2. Lower glume about half the length of the upper glume (3)
_ Lower glume about two-third to three-fourth of the upper one, callus at the bottom of the spikelet sharp, 4–6 mm long, *A. ventricosa*.
3. Each floret disarticulates at maturity, *A. clauda, ssp. clauda*
– Disarticulation occurs only in the lower floret, *A. clauda, ssp. eriantha*.
4. Lemma tips biaristulate (5)
– Lemma tips bidenticulate (9).
5. Panicle mostly flag-shaped, glumes 25–40 mm, individual florets disarticulate, 2–3 mm awl-shaped callus at the base of the dispersal unit, bristles at the lemma tips 8–12 mm, *A. longiglumis*
– Callus blunt, glumes shorter (6).
6. Lemma's bristles 2–5 mm, plants 100–150 cm, $2n = 28$, *A. barbata, ssp. barbata*
_ Lemmas glabrous, florets disarticulate, found in Ethiopia, *ssp. vaviloviana*
_ Florets do not disarticulate, found in Ethiopia, *ssp. abyssinica*
_ Plant shorter than 80 cm, $2n = 14$ (7).
7. Occurring in southeast Spain and Morocco, *A. prostrata*
_ Occurring in Syria and Morocco, *A. damascena*
_ Lemma's bristles 5–12 mm (8).
8. Cultivated form, *A. strigosa, ssp. strigosa*
_ Desert and steppe type, *ssp. wiestii*
_ Mediterranean type, plants 80–100 cm in height, *ssp. hirtula*
_ Disarticulation occurs only at the lower floret, found in Morocco, ssp. *atlantica*.
9. Spikelets small, 1–1.5 cm, known from the Canary Islands, $2n = 14$, *A. canariensis*
– Spikelets of similar size, known from littoral southwest Morocco, $2n = 28$, *A. agadiriana*
– Spikelets larger (10).
10. Awn insertion at the lowest quarter of the lemma, $2n = 28$, *A. murphyi*. Awn insertion at the lower one-third to one-half of the lemma (11)-
11. Spikelets considerably hairy, lemmas of the two lower florets close to each other at their upper part and are nearly parallel, *A. magna*
_ Lemmas of the two lower florets more distant from one another at their top forming V-shaped spikelet (12).
12. Disarticulation scar elliptical oblong, its length about twice the width. $2n = 28$, *A. insularis*
– disarticulation scar elliptical ovate or domesticated form, $2n = 42$, *A. sativa*.
_ Only the lower floret disarticulates, *ssp. sterilis*
_ Each floret disarticulating, ssp. *fatua*
– Domesticated form, ssp. *sativa*.

1.4.1.2 Description:

Section 1. Aristulatae
Series 1. Inaequaliglumes

A. clauda
Plants are short-medium culms 20–70 cm long occasionally masked by taller oat species. Panicle slightly flagged. Glumes unequal, the lower being about half the length of the upper one, $2n = 14$.
ssp. *clauda*
3–5 florets per spikelet, disarticulates at each floret.
ssp. *eriantha*
Two florets per spikelet, disarticulates at the lower floret only.
A. ventricosa
Plants short-medium 25–65 cm. in height, panicle semi-flagged. Glumes unequal in length, the lower one being two-thirds to three-quarters of the upper glume. Spikelet without awns 1.5–2.5 cm, two florets per spikelet only the lower one disarticulates. The callus at the base of the dispersal unit is long, 4–6 mm, and sharp. Awn inserted at the upper two-thirds to three-quarters of the lemma. Lemma tips subulate $2n = 14$.

Series 2. Eubarbatae

A. longiglumis
Plants medium to tall, 40–180 cm in height. Panicle flagged, glumes large and equal. Spikelets made of 2–3 florets, each disarticulates at maturity. Lemma tips aristulate, 8–12 mm long. The callus at the base of the dispersal unit is typically awl-shaped. $2n = 14$. Restricted to sandy and sandy loam soils in desert and mesic habitats in the Mediterranean region.
A. prostrata
Plants short to medium, culms 20–80 cm in height. Panicle is rather dense. Glumes equal or nearly so, spikelet without awns 1.3–3 cm. 2–3 florets per spikelet all disarticulate at maturity. Lemma tips aristulate bristles 3–5 mm long. $2n = 14$. Known from south east Spain and a number of locations in Morocco. Reliable identification of this species may require hybridization with known accessions of this species and cytological examination of the hybrids.
A. damascene
Plants short-medium to tall, culms 20–120 cm in height. Panicles rather dense, glumes equal or nearly so, spikelet small 2–2.6 cm long with three florets disarticulates at maturity, lemma tips aristulate, bristles 3–5 mm long. $2n = 14$, known from one locality in Syria and several locations in Morocco. Reliable identification of this species may require hybridization with known accessions of this species and cytological examination of the hybrids.
A. strigosa
This species has been divided by some taxonomists to several independent species some cultivated and others wild. However, they all are interfertile and are treated

here as one polymorphic species, with four main subspecies, ssp. *strigosa* the domesticated form, ssp. *hirtula* Mediterranean wild type, desert and steppe wild type ssp. *wiestii*, and ssp. *atlantica*, $2n = 14$.

Subspecies *strigosa*

Plants are medium to tall 80–200 cm in height. Panicle rather dense, equilateral, glumes equal or nearly so, spikelets without awns 2–2.5 cm. 1–3 florets per spikelet none disarticulate. Lemma tips aristulate, bristles 6–8 mm long.

Subspecies *hirtula*

Plants are medium to tall, 80–180 cm in height. Panicle is rather dense, equilateral. Glumes equal, or nearly so. Spikelet without awns 2–3 cm, 2–3 florets per spikelet, all disarticulate at maturity. Lemma tips aristulate, bristles 6–10 mm long. It grows on a wide range of soils in the Mediterranean region.

Subspecies *wiestii*

Plants short to medium 40–100 cm in height. Panicle is dense. 2–3 florets per spikelet, bristles 5–8 mm long. Occurs on a wide range of soils in steppe and desert habitats bordering the Mediterranean zone.

Subspecies *atlantica*

This is a recently discovered taxon which was described as an independent species but later appeared to be interfertile with members of *A. strigosa* and accordingly is regarded here as additional ssp. of *A. strigosa*.

Plants short to medium, 40–80 cm. in length. Panicle is rather dense, equilateral. Spikelets, 1.4–2 cm long, 2–3 florets per spikelet, only the lower one disarticulates. Lemma tips aristulate bristles 5–6 mm long. Known from south west Morocco.

A. barbata

This species is similar to the wild forms of *A. strigosa* ssp. *wiesti* and ssp. *hirtula* but can be distinguished from them by its broader panicle with longer internodes and shorter bristles at the lemma tip, and its chromosome number $2n = 28$. It comprises of three main sub species, *barbata, vaviloviana* and *abyssinica*.

Sub species *barbata*

Plant medium to tall, 80–180 cm. in length. Spikelets without awns, 2–2.5 cm long, 2–3 florets per spikelet all disarticulate at maturity. Glumes equal or nearly so, lemma tip aristulate bristles 3–5 mm long. Is an aggressive weed in cultivation and in other man-made habitats. Its similarity to *A. strigosa* ssp. *hirtula* and ssp. *wiestii* has caused confusion between them resulting in considering ssp. *barbata* as having diploid and tetraploid forms and the same for ssp.*wiestii*.

Subspecies *barbata* can easily be distinguished from *A. strigosa* ssp. *hirtula* and *wiestii* by its broader panicle and shorter bristles at the lemma tip, 3–5 mm in ssp. *barbata* and 6–10 in ssp. *hirtula* and *wiestii*. This distinction is sometimes difficult when they grow in mixed populations, because gene flow between them creates intermediate bristle length in up to 10% of the plants or even more.

Subspecies *vaviloviana*

This subspecies is restricted to Ethiopia where it occurs mainly in abandoned fields. It differs from ssp. *barbata* by its glabrous lemmas and occasionally only the upper floret disarticulates.

Subspecies *abyssinica*
This subspecies also confined to Ethiopia where it grows as tolerated weed in barley fields. It does not shatter its seeds at maturity and in that sense can be regarded a domesticated form, but it is never purposely planted. It endures because the Ethiopian farmers cannot select them out by their threshing-winnowing methods. Therefore, this oat is planted, harvested, and consumed together with barley.

Section 2. Denticulatae

A. canariensis
Plants are short to medium, culms 10–80 cm in length. Panicle equilateral, glumes are equal or nearly so. Spikelet of two florets without awns 1.2–1.6 cm long, awn inserted at the middle of the lemma. Disarticulation occurs at the lowest floret only. $2n = 14$, known from the Canary Islands, mainly Fuerteventura and Lanzarote.

A. agadiriana
Plant short to medium 30–150 cm in height, panicle equilateral. Spikelets without awns 1–1.8 cm, with two florets, only the lowest one disarticulates. Lemma tips denticulate. $2n = 28$. This species is known from southwest Morocco. The genetic relations between this species and other oat species have not yet been fully established.

A. insularis
Plants small to medium, 40–90 cm in height. Panicle equilateral. Spikelets v-shaped 1.8–2.5 cm without awns, with 2–3 florets, only the lowest one disarticulates. Lemma tips denticulate. Disarticulation scar elliptical oblong, its length is more than twice its width. $2n = 28$. Known from Sicily and Tunisia.

A. magna
Plants medium to tall 60–120 cm in height. Panicle equilateral, glumes are equal or nearly so. Spikelet without awns 2–3 cm with 2–4 florets only the lowest one disarticulates. Lemma tip dentate, lemmas of the two lower florets are exceptionally hairy and almost touch each other at their upper part. Disarticulation scar is large and elliptic Known from Morocco on heavy crumbling soil. $2n = 28$.
Lately a domesticated form of this species has been developed by transferring the domestication syndrome of the common oat to the wild *A. magna* (see Sect. 2.5.2). Accordingly, the wild form may be regarded as ssp. *magna* and the domesticated type as ssp. *domestica*.

A. murphyi
Plant medium to tall, 60–100 cm in height. Panicle equilateral, slightly flagged. Glumes are equal or nearly so. Spikelet without awns 2–3 cm with 2–4 florets, only the lowest one disarticulates. Awn inserted to the lower quarter of the lemma. Lemmas can be yellow, brown or black, glabrous or hairy. Lemma tips denticulate. $2n = 28$. Known from southern Spain and northern Morocco on heavy crumbling soil.

A. sativa

This is a variable species containing three main subspecies *sativa, sterilis*, and *fatua*. They include domesticated, wild, and weed forms. All are interfertile and share dentate lemma tips and the same chromosome number, $2n = 42$.

Subspecies *sativa*

A domesticated form, extremely variable as result of human selection and breeding. It has been separated to several species by some taxonomists but these are all interfertile. Spikelets with 2–3 florets none disarticulates at maturity. Lemmas are usually yellow and awns may be present, reduced in size or absent.

Sub-species *sterilis*

The most widespread and variable oat type. It is an aggressive weed and common in man-made habitats far from its natural place of origin which is the warm parts of the Mediterranean basin. Plant medium to tall 50–140 cm in height. Panicle equilateral. Spikelet 1.5–4 cm with 2–5 florets, only the lowest disarticulates. Lemma tips denticulate. This subspecies has been divided into a number of species by various taxonomists despite being interfertile.

Subspecies *fatua*

This is mainly a weed form in cultivation and other man-made habitats. Occurs only rarely in primary habitats and usually for a short period. It differs from ssp. *sterilis* mainly by the mode of spikelet disarticulation, which occurs at each floret. It also differs by its geographic distribution which is mainly Europe and North America. This subspecies has also been divided into a number of species by various taxonomists but these are all interfertile.

Chapter 2
My Research Findings in *Avena*

Abstract This chapter occupies the main part of the book. It describes a series of studies undertaken over the past 48 years. During that time a major effort was devoted to series Eubarbatae. In this series, the length of the bristles at the tip of the lemma was used to separate diploids from tetraploids in mixed populations and herbarium material. Accurate identification of the diploid and tetraploid made possible to determine the geographic distribution and the ecological preferences of each of them. Cytogenetic analyses of their triploid hybrids indicated that the diploid form *A. strigosa* is the progenitor of the tetraploid *A. barbata*. It was shown further that this tetraploid emerged through autopolyploidy and its bivalent pairing at meiosis is under control of a single gene. Two species of this series were discovered during the last 50 years. The events leading to their discovery are briefly described. Section Denticulatae contains among others the common oat and its wild forms, all of which share the same genome, commonly designated ACD. This designation is based on comparisons of chromosome morphology and chromosome in situ hybridizations. These types of evidence are in sharp contrast to hybridization-based analyses of chromosome pairing. It was concluded that corroboration of this genome labeling must await the discovery of the diploid progenitors of the polyplois species of section Denticulatae. Over the last 50 years five new species of section Denticulatae were discovered, one of which was the tetraploid progenitor of the common oat and its wild forms. The way by which these five species were discovered is briefly described. The longest project undertaken was an attempt to domesticate the protein-ich wild tetraploid *A. magna*. This species contains accessions with protein content as high as 27 %. While transferral of that protein content to the common oat is impossible because of differences in chromosome numbers, domestication of *A. magna* seemed to be a feasible alternative way to exploit that protein content. After five hybridization cycles between the common oat and *A. magna* we possess domesticated tetraploid lines that are indistinguishable from the common oat and have 25–26 % protein.

G. Ladizinsky, *Studies in Oat Evolution*, SpringerBriefs in Agriculture,
DOI: 10.1007/978-3-642-30547-4_2, © The Author(s) 2012

Keywords Genome analysis · Control of bivalent pairing · Genome identification · Genome labeling · Domestication · Protein content

The botany and the genetics of oat were alien to me when I initiated studying the genus *Avena* in 1964. However oatmeal was a common breakfast at our home. And on our farm oat was grown for hay production, usually in a mixture with vetch. So the subject of my study was not totally away from my heart. Nevertheless, delving into the mystery of this genus and its various species and their peculiarities was a different story.

2.1 Morphology and Ecology of Several *Avena* Species

The first part of my study was devoted to recognizing and becoming acquainted with the diagnostic characters of each species and its range of morphological variation. This was indispensable for identifying each species in its natural habitat and in herbarium material. I must admit that it had taken me some time to become familiar with all the species names in the genus *Avena* and the morphological peculiarities of each of them and to figure out what is a "good" species and what may be regarded as synonym. I had three literature sources for the morphology and the diagnostic characters of the various species: the *Analytical Flora of Israel*, (Zohary and Finebrune1953*), Flore de l'Afrique du Nord* (Maire 1953), and *Wild and Cultivated Oats* (Malzew 1930). The last was one in a series of monographs on cultivated plants and their wild relatives published by several scientists under the leadership of N.I.Vavilov during the 1930s of the twentieth century in USSR.

The next step in my acquaintance with the oat species was to detect them in the field. At that time four wild oat species were recorded in Israel, the diploids *A. wiestii* and *A. longiglumis*, and the tetraploid *A. barbata* and the hexaploid *A. sterilis*. All these species have equal or almost equal glumes. The polyploid species of Israel and in many other countries with Mediterranean climate are aggressive weeds on various soil types. They are common in fields, road sides, and other niches disturbed by man, where they form massive populations. However, they also grow in undisturbed habitats but there they usually form sparse stands. The dispersal unit of *A. sterilis* is the entire spikelet (not including glumes) with large oval disarticulation scar. Lemmas of this taxon terminate in two membranous teeth.

In *A. barbata*, individual florets serve as dispersal units with small disarticulation scar. The lemmas terminate in two bristles of different length.

A. wiestii is similar to *A. barbata* morphologically, and according to the literature differs from it in having lateral tooth at the base of one of the lemma's bristles. In addition, the former is diploid, and the latter is tetraploid. Furthermore, I came across a number of confusing reports of tetraploid *A. wiestii* and diploid *A. barbata*. I then got the feeling that the lateral tooth is not reliable character and cannot be used for classification, and more important that these two taxonomic categories are apparently genetically linked and to prove that I launched a separate study (Ladizinsky and Zohary 1968).

The second diploid oat that was known at that time in Israel was *A. longiglumis*. The dispersal units of this species are the florets and bristles at the lemma tip are 10 mm or even longer. A typical characteristic of this species is its awl-shaped callus at the base of the dispersal unit of 3–4 mm long. Another morphological peculiarity of this species is the unilateral panicle where most of the spikelets are at one side of the panicle axis. *Avena longiglumis* is rather rare species in Israel and throughout its distributional range in the Mediterranean basin, although locally it may be quite common. The reason for this is its affinity for sandy and sandy loam soils.

Once I felt confident about my ability to identify the wild oats in Israel I thought that it will be interesting to examine the *Avena* collection in the Herbarium of the Botany Department at the Hebrew University. This Herbarium possesses large amounts of material collected mainly by Israeli botanists in Israel and in the neighboring countries almost since the foundation of the University.

A herbarium sheet is a sheet of paper to which a pressed and dried plant is glued. A useful herbarium sheet contains the entire plant, if possible, but more important, the organs that enable its accurate identification. Useful specimens also contain written information which includes species name, name of the collector, collecting date, place of collecting, remarks on the habitat where the plant grows such as altitude, longitude latitude, soil type, bedrock formation, and plant community. This information is of great value for evoking primary ideas regarding the geographic distribution of the plant in question and its range of ecological requirements. In addition, it enables workers to return to the same site to collect seeds or conduct ecological and demographic studies.

While I had no difficulty in identifying most of the specimens, one puzzled me. It was collected in 1933 on the sand fields between Amman and Aqaba in Trans Jordan. It was labeled *Avena barbata* by one of the Herbarium staff members. This procedure is common in many herbaria where unidentified material is treated and determined, by expert of the genus in question, sometimes years after the collecting date. The correct name is then inserted into the herbarium sheet together with the name of the person who made the new determination. That specimen looked to me as *A. longiglumis* because of the awl-shaped callus at the lowest tip of the dispersal unit, but it was smaller than the type which I knew from the costal belt of Israel. Then I came across another specimen that was labeled *A. barbata* but in fact was *A. longiglumis*. This specimen was collected at the Muhawat dry water course near the Israeli shore of the Dead Sea. What puzzled me with that specimen was that no sandy soil occurs near the Dead Sea. When I checked on the map the place of origin of that dry water course I realized that it starts about 15 km to the west in a sand plain, east of the town Dimona. I assumed that the *A. longiglumis* seeds were carried by flood water from that plain and established a small population on sandy patch at the bottom of the dry water course.

Winter of 1966/1967 was exceptionally rainy in Israel including the Negev desert and Dimona area. Rain started in November and germination ensued a week or two later. This was the time to visit the site and to look for oat seedlings in the area where I thought *A. longiglumis* of the Dead Sea originated. I uprooted a number of grass seedlings and was able to examine the dispersal units from which they emerged. After less than 15 min I had two seedlings of *A. longiglumis* in my hand which I could identify by the awl-shaped callus at the base of the dispersal unit.

I returned to the same site in March when the oat plants were flowering and was amazed to find a rather dens stand of *A. longiglumis* together with *A. wiestii*. I knew another sabulous habitat of sand fields and sand dunes, between kibutz Revivim and Tse'elim about 25 km west of the original place near Dimona and as expected I found there extensive populations of *A. longiglumis* as well.

This was the first but by no means the last time that I employed ecological extrapolation for expanding the knowledge about the distribution range of a given species not only of *Avena* but also of wild relatives of other crop plants. The procedure is rather simple, when you are searching for sites of a rather rare species, record the ecological components of the site where you have identified that species for the first time and try to find it in another area with similar ecological components.

The morphological difference between the Mediterranean *A. longiglumis* and the desert type which I found in the Negev was minor but the former was larger and more robust than the latter, even if they grew under the same conditions. However, to be sure that they are indeed ecological races and not different species I crossed them with one another. The hybrids developed normally and were fertile, indicating that they are members of the same species. Hybridization tests and examination of the development and fertility of the hybrids have been a standard procedure whenever I found odd types and I was not sure if they were variants of a known species or a new species altogether.

Forty years later, my youngest daughter and her husband established a farm in Revivm area of the Negev where the soil is sand and sand dunes. They raise various types of parrots and exotic fish. I used to come quite often to visit and give them hand in developing the farm. From the first moment there I have realized that their farm is within the potential habitat of the desert form of *A. longiglumis* but I could not find even a single plant of it. There were two possible reasons for that: the area was over grazed by goats and grazing stopped only after the site of the farm was fenced. Also, for their first 2 years on the site there was no rain in the Negev which on average rain fall there is 50 mm. The third year was much better with 35 mm of rain, green patches of grasses started to appear on the sand dunes where the goat grass *Aegilops bicornis* established impressive stands and between them scatter spots of irus, *Iris petrana* with its majestic flowers. While the flowering desert was a wonderful site to behold, more exciting were the few *A. longiglumis* plants growing there. They formed small groups and more often grew individually and quite remote from each other. In later years I saw them regularly when rain fall was sufficient. However, they have appeared more regularly where irrigation was applied. A considerable part of the parrot's diet is based on fresh fruits and to meet that need the family planted various kinds of fruit trees such as oranges, apple, plum, apricot, pomegranate, and guava. Drip irrigation was installed, and in some parts of the orchard small populations of *A. longiglumis* grow along the drip lines every year.

The fruitful survey of the *Avena* material at the Department of Botany suggested to me that misidentification of *Avena* specimen might occur also in herbaria of other countries. To test this I visited the herbaria in Kew Garden and Edinburgh (UK) and Montpellier in France.

The most exciting finding in Kew was several herbarium sheets from Cyprus labeled *A. barbata* but actually were of *A. ventricosa*. This species is characterized by unequal glumes, the lower being two-thirds of the upper one in length, disarticulation takes place only at the base of the lower floret and at the bottom of the dispersal unit there is a long and sharp callus.

The finding of *A. ventricosa* in Cyprus was thrilling because at that time this species was known only in two locations, near Oran in Algeria and Baku in Azerbaijan. While I was checking the herbarium sheets of *A. ventricosa* a man stopped next to my desk. He had seen my name in the visitors book downstairs and had been directed to me. His name was Jack Harlan. It was a complete and most welcome surprise. I knew Harlan by his name and read some of his publications but never expected to meet him particularly in such circumstances. He was on his way home in US from Africa and stopped at Kew Garden to examine some *Sorghum* specimens. I showed him the specimens of *A. ventricosa* which together with this chance meeting had indeed made my day at Kew.

In Montpellier I was mainly interested in the oat material from North Africa. There I saw for the first time *A. mscrostachya* a perennial species endemic to Algeria. Some botanists regard this taxon a member of the genus *Avena* but in my view it is an alien form because it is of perennial, an autotetraploid and self-incompatible, characteristics which do not exist in the annual oats, and is reproductively isolated from them.

Shortly after returning from England and France I traveled to Cyprus to find living plants of *A. ventricosa*. I met there, at the Agricultural Research Institute (ARI) in Nicosia, the person who actually collected the herbarium sheets which I saw in Kew. I found *A. ventricosa* rather common on calcareous soil in stony habitats around Nicosia and even outside the ARI fence. In some places they grew in large and massive populations. It was found in similar habitats near Larnaca and on basalt soils between Larnaca and Nicosia.

As already indicated, when I started my study on *Avena, A. clauda* and *A. eriantha* were not mentioned in the Analytical Flora of Israel, nor was either of them in the herbarium of the Botany Department. However, in one of my excursions aimed at getting a better idea of the ecology of the oat species native to Israel, I came across a small stand of short stature oat plants with unequal glumes and dispersal unit made of the two florets. In other words it was *A. eriantha*, formerly known as *A. pilosa*. Later I found the *A. clauda* type in which the floret is the dispersal unit. Occasionally these two taxa grow in pure stand but more often side by side in the same population. Their populations are usually composed of a small number of plants that thrive on shallow calcareous and terra rossa soils in Mediterranean and steppe vegetation. Both types are early maturing, about 2 weeks before *A. sterilis* and *A. barbata* which are common in places where *A. clauda* and *A. eriantha* grow. Thus, when the polyploidy oats flower they mask the diploid species which have already dispersed their seeds.

After finding these two oat types in Israel I had the opportunity of crossing them with one another and to examine their hybrids. As already mentioned, these hybrids developed normally, where fertile and disarticulation occurred at the lower floret only. In F2, segregation to the two fruiting type had taken place indicating that this character is controlled by two complementary genes and the clauda morph being recessive to that of eriantha (Ladizinsky 1968; Ladizinsky and Zohary 1971).

2.1.1 Performing Crosses in Avena

From the beginning of my study of the genus *Avena* it was clear to me that for establishing the identity of the various species and the difference between them and other species, it would be necessary to perform hybridization experiments on a large scale. Although the oat flower is relatively large, making crosses and getting hybrid seeds is not simple. To be successful one must be familiar with flowering pattern of the oat panicle and the exact time when the stigma is receptive and the anthers mature.

Flowering in oat panicles goes from top to bottom and in each spikelet the lower floret flowers a day before the upper one, the two florets of the spikelet rarely flower on the same day. The first to flower is lower floret of the upper most spikelet. The next day the upper floret of the same spikelet will flower as well as the lower floret of the lower spikelet. On the third day the upper floret of the second spikelet will flower. On the same day, flowering will take place in the lower floret of the spikelet connected to the upper panicle node by the longest pedicel. In the fourth day the upper floret of the latter spikelet will flower as well as the lower floret of the spikelet connected to the same node by shorter pedicel and so on (Fig. 2.1). As flowering proceeds the order is retained but tracing the florets that are going to flower on a given day cannot always be accurately done. When crossing florets of panicles that have already started flowering, the identification of florets that are supposed to reach anthesis on the same day can be made by selecting spikelets in which the lower floret has already been flowered and therefore has no anthers, and then working with the upper flower which still possesses anthers. In this way the flowering rhyme of the panicle can be followed.

As in other grasses, flowering time in oats is short and its onset takes place at a specific time of the day. In my place it starts at about 4 pm and lasts for about 30 min. The flowering starts with swelling of the lodicules which push the lemma and the palea to opposite directions. Then the stigma is exposed and at the same time rapid elongation of the anthers filaments occur. Shortly after that the anthers burst along existing slits and release the pollen. Hybrid seeds in oats can be obtained only if the cross is made at the short period of the natural anthesis.

In oats as in other plants, artificial crosses are made of two stages, the emasculation, removal the anthers of the floret that is going to be crossed, and pollination. These two stages can be made on the same operation or the emasculation can start a bit earlier, particularly if a large number of hybrid seeds are required.

Fig. 2.1 Flowering pattern
in oats (see text)

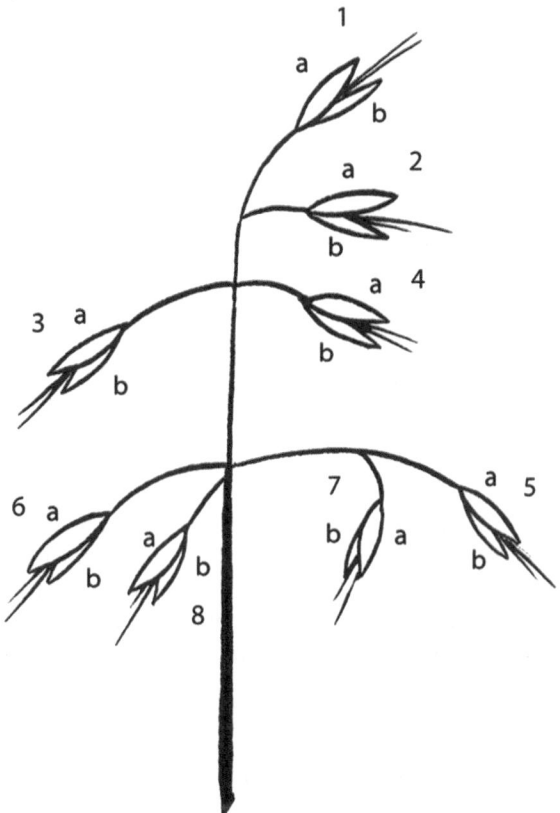

2.2 Series Eubarbatae

I have adopted Malzew's terminology for this group of diploid and tetraploid oat species but developed my own concept of their evolution. This group contains wild and weed forms, domesticated and semi domesticated types, all having basically similar morphology with more or less equal glumes, and florets that serve as dispersal units.

When I started studying this group, there was no published material on morphological characters that distinguish between diploids and tetraploids. Furthermore, I came across a number of reports in the literature of diploid and tetraploid forms for the species *A. wiestii* and *A. barbata*. In an attempt to solve this enigma I selected a place about 15 km north of Be'er Sheva in a hilly area where diploid and tetraploid of *A. wiestii* were reported. My idea was to obtain the chromosome number of individual living plants, to check carefully their morphology and to find out if there is any morphological trait associated with $2n = 14$ or 28, the diploid and the tetraploid chromosome numbers.

Common chromosome counts in plants are usually done in root tips of germinating seeds. A more laborious method is to check chromosome number in embryonic inflorescence, at meiotic division in pollen mother cells (PMC) from which pollen grains are produced. In oats this division is taking place in tillers at the flag leaf stage where the embryonic panicle is developing. Some experience is necessary to correlate the time of meiosis with the length of the tiller that terminates in a flag leaf. The meiotic division itself takes place in two stages, the one following continuously on the other. Both stages comprise the substages of prophase, metaphase, anaphase, and telophase. For the purpose of chromosome counts metaphase of stage one is the most useful but it is the shortest of all other substages and detecting it can be laborious and tedious work. However, the advantage of this particular substage is that the chromosomes are arranged in pairs and the number of the elements to be counted is half of the somatic chromosome number.

When I arrived at the site where plants of the Eubarbatae group were rather common I started collecting flag leaf tillers and developed inflorescences from individual plants. Altogether I collected material from about 30 different plants before returning to the lab to check the chromosome numbers. I was pleased to find that some of the plants, in which I had good view of the meiosis were diploids ($2n = 14$) and some tetraploids ($2n = 28$). On comparing the morphology of these two types I observed differences in two characters: panicle shape and length of bristles at the tip of the lemmas. When I put the panicles of the tetraploid and the diploid side by side it became obvious that the former was usually larger and more open than the latter. The diploids had characteristically smaller spikelets, shorter internodes, and more spikelets per peduncle making the panicle denser compared to the tetraploid panicle (Fig. 2.2a, b). More remarkable was the difference in the bristle length between the two types: 6–9 mm in the diploids and 3–5 mm in the tetraploids (Fig. 2.3a, b), Table 2.1.). The bristles of the diploids were usually purple but this character alone was insufficient for indicating the ploidy level.

With that information, I returned to the site and begun to examine the Eubarbatae plants according to their panicle shape and bristle length. I was surprised how easy it was. From plants identifiably diploid or tetraploid, I collected tillers for chromosome counts and was pleased to find nearly complete agreement between the morphological and the cytological results. There remained, however, a small number of plants whose ploidy was not sure because their bristles were of intermediate length. Later, I noticed that phenomenon only where diploids and tetraploids were growing side by side which suggested to me the occurrence of gene flow between the two.

2.2.1 The Diploids

Once it became possible to separate, on morphological ground, diploids from tetraploids I started exploring the distribution of the diploids in Israel. Three types were found which can be regarded as ecological races. The first one is a desert, or steppe type which is common in the southern part of the country, where annual rainfall is between 100 and 250 mm, on several soil types from sand to loess. The second one occurs along the costal belt on sandy and sandy loam soils with annual

Fig. 2.2 Panicles of **a** *A. strigosa* ssp. *wiestii* ($2n = 14$), **b** ssp. *barbata* ($2n = 28$)

rainfall of about 500 mm, and is characterized by extremely long bristles, up to 12 mm. The third one was found on calcareous bedrock in Judea foothills, on mount Gilboa, and on basalt soil in the Golan Heights where rainfall is 500–700 mm.

While the desert type is commonly referred to as *A. wiestii* (ssp. *wiestii*) I considered the other two types to be *A. hirtula* (ssp. *hirtula*).

To be certain that these three types are members of the same species I crossed them with one another. The hybrids were normal and fertile, meaning that they should be regarded as races or subspecies of the same species.

Furthermore, when I had examined bristle length of herbarium material deposited in Kew Garden, Edinburgh, and Montpellier, I could easily recognize diploids that

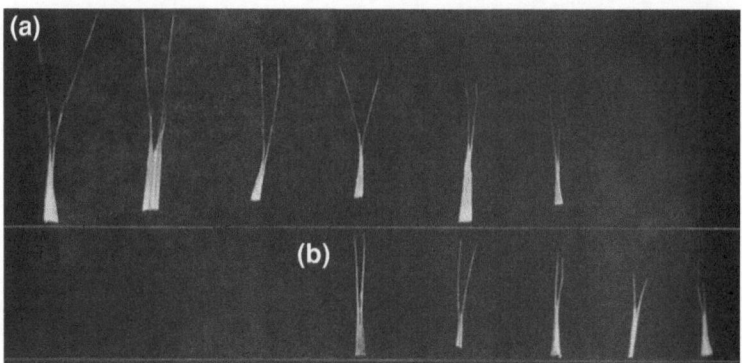

Fig. 2.3 Variation in bristle length at the tip of the lemma in mixed stand of **a** *A. strigosa* ssp. *hirtula* ($2n = 14$) and **b** *A. barbata* ($2n = 28$)

Table 2.1 Variation in bristle length in mixed stand of *A. barbata* ($2n = 28$) and *A. strigosa* ssp. *hirtula* ($2n = 14$)

	Bristle length in mm										Total
	2	3	4	5	6	7	8	9	10	11	
ssp. hirtula					3	7	8	7	3	2	30
A. barbata	2	6	8	6	4	1					27

were determined as *A. barbata*. On the basis of that survey I visited populations of ssp. *hirtula* in Turkey, Cyprus, Crete, and the Iberian Peninsula where I could corroborate the ecological requirements of this diploid oat and its spatial contact with the tetraploid *A. barbata*. Hybrids that I made between representatives of ssp. *hirtula* from different parts of its wide distribution range were viable and fertile, indicating that it is a coherent species with a well-defined genome. Ecologically it is quite diverse and occurs in desert and steppe vegetation as well as in Mediterranean climate. It is usually confined to primary habitats but also grows in disturbed habitats and as weed in cultivation, particularly in the Iberian Peninsula.

This species has a domesticated derivative commonly known as *A. strigosa*. Its main feature is the non-shattering seed phenotype, meaning that at maturity the florets remain attached to the spikelet axis, a character controlled by two genes. Another domesticated form in this group is *A. nuda* with naked seeds. The interfertility between *A. strigosa, A. wiesti*, and *A. hirtula* indicated to me that the taxonomic status of these taxa should be reduced to subspecies.

2.2.2 The Tetraploids

The tetraploid forms of series Eubarbatae are by far more common than the diploids. The dominant type in this group is ssp. *barbata* of *A. barbata* which is a common weed in the Mediterranean region and in Mediterranean-like climates and thrives in a range of soils and altitudes, from sea level up to 1500 m or even more. It manifests little morphological variation, mainly in spikelet size and lemma color.

Besides *A. barbata,* series Eubarbata contains another two tetraploids ssp. *abyssinica* and ssp. *vaviloviana* which are semi-domesticated forms and occur almost exclusively in the Ethiopian heights. Morphologically the *abyssinica* form is characterized by seed nonshattering, whereas in the *vaviliviana* form the individual florets disarticulate (though occasionally the lower floret remains attached to the spikelet axis).

When I read the description of these two taxa I wondered if these Ethiopian oats were domesticated forms, and if not, how they distribute their seeds and what their habitat was. The best way to answer these questions was to go to Ethiopia and check these oat forms in their natural places.

In Ethiopia it became apparent that none of these oat species is purposely grown by the farmers as a crop. Furthermore, ssp. *abyssinica* was much more common than ssp. *vaviloviana* and both were confined to arable land and never were found in primary habitats. Occasionally, ssp. *barbata* was also found next to these oats especially with the *vaviloviana* form, mainly in abandoned fields.

Subspecies *abyssinica* is found, almost exclusively, in barley fields. At the vegetative stage it is difficult to tell if a given plant is oat or barley. It can be achieved only with the emergence of the inflorescence, a spike in barley and a panicle in oat. Weeding out the oat plants at this stage is problematic because it might harm the adjacent barley plants. Nevertheless, in rainy years some weeding is done but not so in dry years so as not to lessen the yield. So, it is regularly harvested and threshed with the barley crop. With the threshing and winnowing methods used by the Ethiopian farmers it is impossible to separate the ssp. *abyssinica* seed from barley because they have the same shape and weight. As a result, seeds of this subspecies are common in barley samples in the Ethiopian rural markets and the people use this oat for whatever purpose they use barley, especially for brewing bear. Naturally, ssp. *abyssinica* seeds occur also in the barley sowing stock. By definition, this oat is a domesticated form but it is not a crop and can be regarded as tolerated weed.

Of all oat taxa, only the tetraploid forms of series Eubarbatae are found in Ethiopia and apparently grow there for many years. In the last few decades, however, the hexaploid ssp. *sterilis* can be found there but it is apparently a recent introduction as contaminant in imported wheat seeds. It seems logical to assume that in historical times *A. barbata* seeds arrived in Ethiopia as contaminant in crop seeds from the Middle East. After all, most grain crops of Ethiopia originated from the Middle East: wheat, barley, pea, chickpea, lentil, faba bean ,and bitter vetch. They are grown at high altitudes, 2000–3000 m above sea level where the temperature is favorable for these crops and also for *A. barbata*. Evidently the farmer's field had been an appropriate habitat for *A. barbata* in Ethiopia but not so for other ecological niches to which this species is not adapted.

The genetic difference in seed dispersal pattern between ssp. *barbata* and *ssp. abyssinica* is important for understanding how the latter evolved. Hybrids between the two disperse their seeds as ssp. *barbata* indicating the recessive nature of the non-shattering ssp. *abyssinica* seeds. Segregation in F2 revealed that this character is governed by four genes (Jones 1940) and two genes in two *A. abyssinica* x *A. barbata* hybrid combinations that I tested (Ladizinsky 1975b).

Cytologically, the tetraploid forms of series Eubarbatae are regarded as allo-ployploids. Their 28 chromosomes are arranged at meiosis in 14 pairs (bivalents). As all other annual oat species they are selfers and anthesis usually occurs while the florets are still close or right after flowering. However, when individuals from different *A. barbata* populations and even from the same populations are crossed with one another, meiosis of the hybrids is irregular. Instead of 14 bivalent some of the chromosomes are left unpaired (univalents) or associate in groups of three (trivalents) or four (quadrivalents) or even groups of five and six chromosomes (higher multivalents). The formation of multivalents of different sizes in hybrids between different lines indicate massive chromosomal rearrangements in the form of translocations which have taken place at the tetraploid level, and it is in sharp contrast to the diploids which are cytologically highly uniform. The common outcome of the irregular meiosis is reduction of pollen fertility and seed set. Nevertheless, I found that hybrids with a single rearrangement were almost as fertile as their parents and when the parental lines differ by three rearrangements seed set was about 15 % of the parental lines. The significance of the chromosomal flexibility in *A. barbata* and its only moderate effect on fertility will become clear after we examine the behavior of the hybrids between the diploid and tetraploid forms of series Eubarbatae.

The intimate relation between the diploids and the tetraploids has been indicated above in terms of their similar morphology, but how close are they genetically? Was the diploid involved in the formation of the tetraploid, and if so, which other diploid participated in that process?

The appropriate way for assessing genetic relationships between species is to try to cross them with one another and to see the outcome in terms of chromosome behavior at meiosis and fertility. If there is any peculiarity in this case it usually would be revealed by the chromosome behavior at meiosis of the hybrid. Within species, chromosome number is dual and each chromosome has its homolog which possesses the same basic genetic information. Other pairs contain different information and are nonhomologous to each other. At meiosis homologous chromosomes pair with one another and appear as bivalents. In such bivalents, attachment between the two homologs can be confined to one chromosome arm or to both arms. Accordingly, the bivalent is shaped either like a rod or like a ring (Fig. 2.4a, b). The point of attachment between pair of homologous chromosome is called chiasma (pl. chiasmata). When no chiasma occurs between homologous chromosomes they would remain unpaired and at Anaphase I, unlike paired chromosomes, they would not move to the poles as other chromosomes. As a result, the daughter cells will miss one chromosome or may get an extra one and in either case will usually abort.

In an interspecific hybrid between species that are genetically remote from one another the chromosomes may not pair at meiosis, would stay as univalents, and

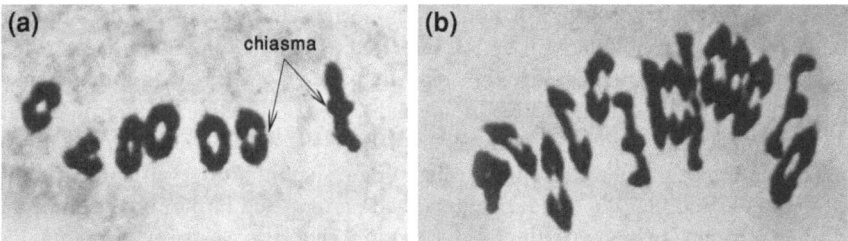

Fig. 2.4 Metaphase I chromosomes, **a** six ring and one rod bivalents of *A. strigosa*, **b** fourteen bivalents of *A. barbata*

the hybrid will be sterile. When the chromosomes of the parental species are more related some bivalent and multivalent associations might be formed, but in most cases the hybrid will similarly be sterile. Thus, the pattern of chromosome associations in the interspecific hybrids can be used in assessing the genetic similarity between the parental species. Genetic similarity can be more reliably expressed by the average number of chiasmata per cell because this index indicates the chromosome segments in the parental species which are sufficiently similar for allowing chromosomes to pair and hence chiasma formation.

2.2.3 Production of Triploid Hybrids in Series Eubarbatae, Their Cytology, and Fertility

When crossing species of different polidy levels, it is common to use the species with higher ploidy level as the female parent and this is what I did. I used a number of tetraploid and diploid lines and tested about ten different triploid combinations. Crossability between the parental lines was good and about one seed was obtained out of two cross. Later, I discovered that triploid hybrid seeds could also be obtained with the diploid serving as female parent, but in that case considerably fewer seeds were obtained.

The triploid hybrids developed normally but were partially sterile. Pollen fertility ranged between 25 and 42 % and seed set ranged between 0.7 and 4 % in bagged panicles and 5–16 % in open, non-bagged panicles. This came as a great surprise because triploid hybrids are usually expected to be sterile and a dead end as far as gene flow between the parental species is considered.

No less interesting was the cytology of the triploid hybrids and the pairing pattern of their chromosomes at metaphase I of meiosis. In hybrids between a diploid and a tetraploid species two extreme cases may occur that the diploid genome is not related to any of the two tetraploid genomes, or the diploid genome is present in the tetraploid species because the diploid species participated in the formation of the tetraploid. In the first case all or most of the chromosomes will appear as univalents at meiosis. In the second case the chromosomes of the diploid

will pair with their homologs or homeologs in the tetraploid, while the chromosomes of the other genome will be left unpaired as univalents. The pioneer wheat
cytogeneticist Hitoshi Kihara used this pairing pattern to identify diploid genomes
in teteraploid species in the wheat group and termed this experimental procedure
"genome analysis" (Lilienfeld and Kihara 1951).

The pattern of the chromosome pairing at meiosis that I observed in the triploid
hybrids indicated that the diploid genome had participated in the formation of the
tetraploid forms. In many cells the association pattern was seven bivalents and
seven univalents (Fig. 2.5). In other cells the number of univalents ranged between
4 and 9 and the bivalents from 3 to 8. In these cells multivalents were common,
mainly trivalents (0–2) or a quadrivalent. These multivalents are probably the
result of chromosomal repatterning in the tetraploid lines as already mentioned.
Another indication for the intimate genetic relations between the diploid genome
and one of the tetraploid genomes is the number of chiasmata per chromosome in
the triploid hybrids. It was 1.92−1.80 in the parental lines and about two-thirds of
that (1.25–1.33) in the triploids. The reduced number of chiasmata per chromosome resulted from the occurrence of univalents and multivalents. Even a stronger
indication came from the similar number of chiasmata per cell in the diploids and
the triploids (12–14)

2.2.4 Restoration of Fertility and Stabilization
of Chromosome Numbers in F2

Forty-four F2 plants originating from bagged and unbagged panicles of the triploids
were examined for their $2n$ numbers, chromosome pairing pattern at meiosis and
fertility. The results were similar for the two groups. Thirty-three plants had
chromosome numbers near or the exact tetraploid $2n$ number (26–28). In two F2
derivatives with $2n = 28$ the chromosome were arranged in 14 bivalents and their
fertility was over 90 %. The fertility of those with 27 and 26 chromosomes was in
the range of 20–80 %.Two plants had the diploid chromosome number and in one
of them the chromosomes were arranged as seven bivalents and in the second one a
trivalent and univalent was occasionally formed. This last plant, however, was more
fertile than that with the perfect meiosis. The rest of the plants had 22–25 chromosomes and were the least fertile, with a fertility rate similar to the triploid plants.

The range of chromosome numbers and the high fertility of those derivatives
with euploid chromosome numbers indicate that between diploid and tetraploid
forms in Eubarbatae group exists quick and almost unnoticeable gene flow via a
triploid bridge. Some derivatives had already acquired the euploid chromosome
number in the F2 and were highly fertile. Moreover, plants with aneuploid numbers were sufficiently fertile to produce euploid plants in the F3 generation. Gene
flow between diploids and tetraploids in series Eubarbatae goes in both directions,
from diploid to tetraploids and vice versa. The first is the more common and
apparently the more effective because unlike the diploids, the tetraploid forms are

Fig. 2.5 Seven bivalents and
seven univalents in *A.*
barbata x *A. strigosa* triploid
hybrid

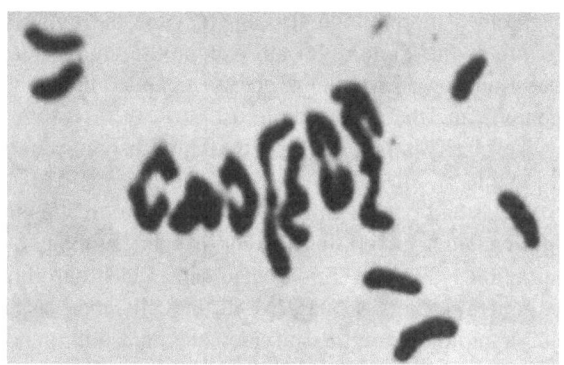

capable of storing a larger amount of variability with minor harm due to missing a
particular chromosome or chromosome segments. Through the triploid bridge, the
tetraploid can absorb the genetic variability of isolated diploid populations, each of
them adapted to different ecological conditions. This may be one of the reasons
that the tetraploids are much more common and aggressive than the diploids. On
the other hand, some diversity of the tetraploids can be transmitted to the diploids
but its integration in the diploid genome is more difficult because it may upset the
genetic balances which promote adaptation to the diploid ecological niche.

Finally, early in this chapter it was described how I managed to separate diploids
from tetraploids according to the length of the bristles at the tip of the lemmas in
places where they grew side by side. It was also indicated that certain proportion of
the plants in such places, mainly with $2n = 28$, exhibit intermediate bristle length.
This overlap can be explained now in terms of gene flow via triploid bridges.

2.2.5 The Genomes of the Tetraploids in Series Eubarbatae

The morphological and cytogenetic evidences support the diploid forms of series
Eubarbatae as one of the parents of the tetraploid forms. In other words, one of the
tetraploids genomes was contributed by the diploid forms of this series. What then
is the identity of the other diploid which contributed the second genome to the
tetraploid forms? This question is legitimate because the tetraploid forms cytogenetically behave as allotetraploids, but as shown later, other factors may be
responsible to that allotetraploid-like cytological performance.

Of the diploid species which were known at that time, *A. clauda, A. ventricosa*
and *A. longiglumis*, the former two were cross-incompatible with the tetraploid
forms and the hybrids with *A. longiglumis* exhibited irregular meiosis with many
multivalent associations and were completely sterile. Furthermore, morphologically allotetraploids usually manifest combined morphology of their diploid parents. None of these diploid characters can be identified in the tetraploid forms of
series Eubarbatae.

From a morphological point of view the range variation of the tetraploids is found in the diploid forms, except the shorter bristles at the lemma tip. The question however is the degree of genetic affinity between the tetraploids second genome and that of the diploids.

To clarify this I doubled the chromosome number of the domesticated form known as *A. strigosa* (no. 6516) and line no. 6541 of ssp. *hirtula*. These autotetraploids had typical chromosome association at meiosis with the formation of 2–5 multivalents per cell. In the hybrids between the two autotetraploids and the native tetraploids of series Eubarbatae (no. 6533) chromosome association was nearly complete in the vast majority of the cells and was similar to that observed in the autotetraploid parents. This indicates that both genomes of the native tetraploids are similar to that of the diploids in that series.

2.2.6 Genetic Control of Bivalent Pairing in A. babrbata

To resolve the discrepancy between the mode of chromosome association in the native tetraploids and the artificially produced autotetraploid, and to check the possible involvement of genetic factors here, the pattern of chromosome association examined in 133 F2 derivatives of F1 hybrids involving the two autotetraploid lines and the same *A. barbata* accession. In the 6533 × 6516 F2 family, 2 of the 63 plants had only bivalent pairing while in the 6533 × 6541 family in 21 of the 70 derivatives 14 bivalent association were observed in each cell. The rest of the F2 plants exhibited various numbers of multivalents in each cell (Ladizinsky 1973).

The segregation pattern in both families indicated the control of a single gene on bivalent pairing. However, in the first family the pattern is of tetrasomic inheritance but disomic in the second generation. According to these results, the bivalent pairing in the native tetraploids is due to genetic control of a single gene in a recessive state. A former indication of this was the paring pattern in one of the F2 progeny of the hybrids between diploids and tetraploids of series Eubarbatae. This particular plant had 26 chromosomes and most of them were associated as multivalents. Apparently the missing chromosomes were the carriers of the gene for bivalent pairing.

The first report that bivalent pairing in allotetraploid species is due to genetic factor was obtained by examining the pattern of chromosome association in a monosomic line of chromosome 5B in wheat (Riley and Chapman 1958). As far as I know, it is only in the tetraploid forms of Eubarbatae the preferential pairing has been related to a single gene.

2.2.7 The Possible Origin of the Gene for Bivalent
Pairing in A. barbata

The discovery that a single gene is controlling the bivalent pairing in the tetraploid forms still does not provide any clue regarding its actual action, or how it had

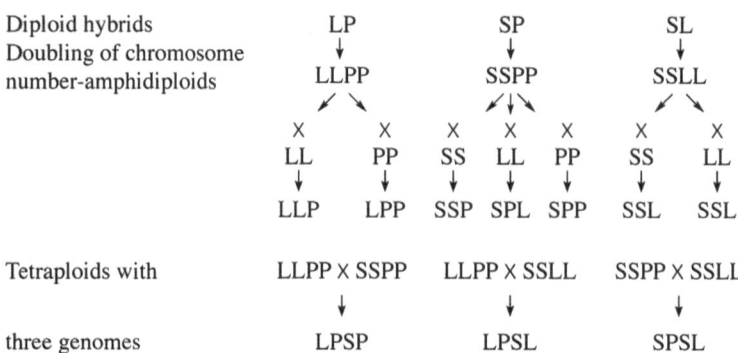

Diploid hybrids	LP		SP			SL	
Doubling of chromosome	↓		↓			↓	
number-amphidiploids	LLPP		SSPP			SSLL	
	↗ ↘		↗ ↑ ↘			↗ ↘	
	X X		X X X			X X	
	LL PP		SS LL PP			SS LL	
	↓ ↓		↓ ↓ ↓			↓ ↓	
	LLP LPP		SSP SPL SPP			SSL SSL	

Tetraploids with	LLPP X SSPP	LLPP X SSLL	SSPP X SSLL
	↓	↓	↓
three genomes	LPSP	LPSL	SPSL

Fig. 2.6 Hybrid combinations between *A. strigosa* (S), *A. prostrara* (P), *A. longiglumis* (L) and their artificially produced amphidiploids

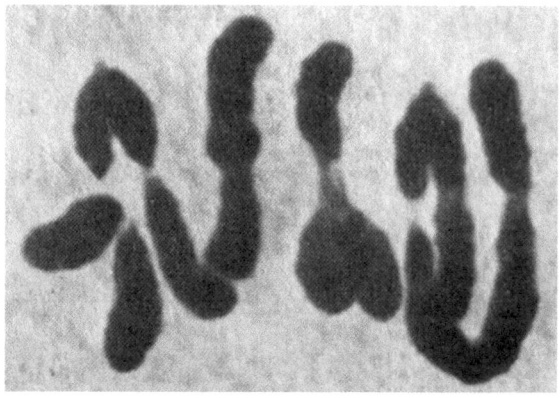

Fig. 2.7 A univalent, bivalent, two trivalents, and pentavalent in *A. strigosa* x *A. longiglumis* hybrid

evolved. Was it formed after the doubling of the chromosome number of the diploid or has it pre-existed in the diploid forms of series Eubarbatae.

To examine the possibility of the pre-existence of such a gene at the diploid level I artificially created interspecific diploid hybrids in which a given genome was presented in a single dose with the presence of another genome. Another combination was produced after creating autotetraploid of a given species that was crossed with another diploid. In that triploid hybrid two genomes were identical together with alien genome (Ladizinsky 1974). Altogether, three diploid hybrid combinations were examined, three artificially produced allotetraplois, seven triploid combinations, and three tetraploid hybrids containing three genomes (Fig. 2.6).

The three species involved in that experiment were *A. longiglumis* (L), *A. strigosa* (S), and *A. prostrate* (P) a diploid species of series Eubarbatae that I discovered in Spain in 1969 (see Sect. 2.3.9).

The chromosomes of *A. longiglumis* and *A. prostrata* paired fairly regularly in the PL interspecific diploid hybrid. Most of the chromosomes were associated as

Table 2.2 Meiosis in parental species and their hybrids and in triploid hybrids with different genomic combinations

Parents and hybrids	2n	No. of cells	I	II Rod	II Ring	III	IV	V	VI–VIII
A. longiglumis (LL)	14	30	–	1.13 (0–2)	5.87 (5–7)				
A. prostrata (PP)	14	30	–	1.07 (0–3)	5.93 (4–7)				
A. strigosa (SS)	14	30	–	1.1 (0–3)	5.9 (4–7)				
LP	14	100	2.1 (0–6)	3.06 (1–6)	2.68 (1–4)	0.03 (0–1)	0.06 (0–1)		
LS	14	150	1.32 (0–4)	1.52 (0–3)	0.06 (0–2)	1.7 (0–3)	0.42 (0–2)	0.43 (0–2)	0.09 (0–1)
SP	14	100	0.12 (0–1)	0.34 (0–2)	0.16 (0–1)	1.78 (0–3)	0.24 (0–2)	0.8 (0–2)	0.42 (0–2)
LLP	21	120	3.63 (1–6)	1 (0–2)	2.95 (1–5)	3.09 (1–6)	0.06 (0–1)		
LPP	21	100	4.36 (2–7)	1.08 (0–3)	2.82 (0–5)	2.96 (1–5)			
SSP	21	125	4.71 (2–6)	2.14 (9–5)	2.81 (0–7)	1.21 (0–3)	0.43 (0–2)	0.19 (0–1)	
SPP	21	120	4.32 (1–6)	1 (0–3)	3.52 (2–7)	2.22 (0–4)	1.17 (0–1)	0.05 (0–1)	
SSL	21	120	3.26 (2–6)	1.53 (0–4)	3.52 (1–5)	2.01 (1–4)	0.22 (0–1)	0.06 (0–1)	
SLL	21	120	2.53 (0–5)	1.02 (0–3)	1.93 (0–3)	2.56 (0–5)	0.73 (0–2)	0.26 (0–2)	0.09 (0–2)
SPL	21	150	2.8 (1–5)	1.64 (0–5)	0.76 (0–4)	2.03 (0–5)	0.9 (0–3)	0.35 (0–2)	0.3 (0–2)

bivalents but the number of the rod bivalents was almost three times higher compared to the parental species. In addition, occasional univalents and a trivalent or a quadrivalent were observed. Despite the fairly regular chromosome pairing, the hybrids produced only a small number of seeds.

In contrast, in the SL and SP diploid hybrid chromosomes were associated in complex multivalents (Fig. 2.7) indicating that the parental species diverged by five chromosomal rearrangements (Table 2.2).

Chromosome association in seven triploids with all possible genomic combinations was interesting. The triploid SPL exhibited the most complex chromosome association and the greatest number of chaotic multivalents. Triploids with the most regular chromosomal association were observed in the SSP and PPS genomic formulas in which cells with seven bivalents were quite frequent. This indicates that the accessions of *A. strigosa* and *A. prostrata* contained factors which promote preferential pairing in the presence of other genomes.

The role of such a factor, or factors, at the diploid level is puzzling because each chromosome there is presented by two homologs which regularly pair with one another at meiosis. It is possible that these factors have other function at the diploid level but manifests preferential pairing in unusual circumstances as in tested triploid hybrids. It is also not yet clear if that factor for preferential pairing is related in any way to the gene controlling bivalent pairing in *A. barbata*.

2.2.8 The Genome Labeling of the Tetraploid A. barabata

A genome is the chromosome set of the gamete. A diploid species contains two identical genomes, an autotetraploid four sets of the same genome and allotetraploid two sets of two different genomes. In diploid *Avena* species with x = 7 each genome contains seven chromosomes which by definition are species specific. Unfortunately, some authors have a different view on the genome issue. For them, several oat species share the same genome. This is not only confusing but also misleading in the discussion of the identity of the polyploid oat forms.

Nishiyama (1929) was the first to suggest that the genome of *A strigosa* designated by him as the A genome is present in *A. barbata*. Accordingly, he proposed the genomes of this tetraploid AABB but the donor of the B genome was not clear at that time. In 1936 after examining the hybrid between *A. barbata* and the autotetraploid *A. strigosa* he noticed the great similarity between the two genomes of *A. barbata* and their resemblance to the A genome. I became aware of Nishiyama's work after conducting a similar study and reached the same conclusions but with one difference. I regarded the genomic formula of *A. barbata* as AA A'A' which better indicates the close affinity between the two genomes (Ladizinsky 1968; Ladizinsky and Zohary 1968).

Support to the idea that the *A. barbata* group evolved essentially via autopolyploidy has been provided by DNA:DNA in situ hybridization, using total genomic DNA of *A. strigosa* as a probe on chromosomes of ssp. *vaviloviana*. All the chromosomes of the tetraploid were labeled strongly and uniformly by *A. strigosa* DNA (Katsiotis et al. 1997). However, one has to bear in mind that this experimental technique is inadequate for separating the various A genome types, original or modified.

Despite all the evidence accumulated on the nearly autotetraploid origin of *A. barbata* group, the genome of that group is still designated as AABB. I wonder if the authors using this formula know something that I donot know, or are simply not aware of the evidence contradicting that formula, or perhaps do it habitually by copying it from older publications. I believe the time has come for oat specialists to accept the accumulated evidence regarding the autotetraploid origin of the *A. barabata* group.

2.2.9 Recently Described Species in Series Eubarbatae

In the last 40 years the genus *Avena* has been enriched by several species that were not previously known. Some of them are of series Eubarbatae.

2.2.9.1 Avena prostrata

In 1969 I visited the Iberian Peninsula for a field study of wild oats there. My main interest was the distribution and ecology of the diploid and tetraploid forms of the Eubarbatae series. However, I came across all other species which were listed for that territory with which I was familiar from my study in Israel.

To be successful, a mission like that requires careful preparation. Three elements are of prime importance, the geology of the area, soil types, and vegetation. In addition, information on the target species in the territory that are going to be explored must be adequate and reliable. Sources are geological and soil maps, and publications and herbarium material of the target species in that area.

The general route of the excursion in Spain included the plateau around Madrid, Madrid—Valencia road, down to Alicante, Murcia, Almeria, Granada, Malaga, southern tip of Spain, Seville, then to Faro, Lisbon, Coimbra in Portugal, back to Spain in Salamanca, Valladolid, and back to Madrid. From that main route I used to take detours to examine unique ecological niches as I expected them of different rock and soil formations. From my previous field excursions I have learned that these two elements dictate the landscape and the vegetation. Whenever I notice different bedrock formations or soil types I stopped to look at the wild oats and to collect seeds.

The tetraploid *A. barbata* and the wild forms of *A. strigosa* are common in the Iberian Peninsula and occasionally grow in mixed stands, particularly in ecological niches which are favorable by the diploids. Using the bristle length at the tip of the lemma and panicle shape I had no difficulty in separating the two. I found the diploids mainly on sandy soil along the coastal areas. They had very long bristles at the tip of the lemma and occasionally grew together with *A. longiglumis*. This type was common in the Seville area and along the Seville-Faro road. The other ecological race was found in Portugal between Ameixial and Ervidel in association with the cork oak. The wild *A. strigosa* I found there was in primary and secondary habitats and was also common along the road from Coimbra to Salamanca.

In general, the situation in the Iberian Peninsula was quite similar to what I had observed in Israel. However, on the plateau I found wild *A. strigosa* as weed in cultivation, particularly in the Salamanca area. Another difference is that along most of the route I have not come across real desert vegetation, where I expected to find the desert race of *A. strigosa*. I hoped to find it in the south eastern corner of Spain. This area is under the rain shadow of the Sierra Nevada with an annual rainfall of 200–300 mm.

From Lorca I took the road to Aguilas. That road runs through an irrigated valley and then up the metamorphic hills. The contrast between the vegetation in the valley and on the hills was dramatic. Metamorphic bed rock creates a dry habitat which is exacerbated by the low rainfall in that region.

In the valley I found only *A. barbata* and ssp. *sterilis* but on the hills I found small patches of small stature oat plants with panicle which seemed to me close to ssp. *wiestii* but with somewhat shorter bristles. There were also a few *A. barbata* plants there which I could identify rather easily. I came across this type again and again in the same bed rock formation and collected seeds from them.

Back home I germinated the material I brought from the Iberian Peninsula. The plants which arose from the seeds I collected on Lorca-Aguilas road and from other locations in that general area with the same bed rock formation had a marked prostrate growth habit. I was wondering if it was a new type of wild oat or a variant of the *A. strigosa* desert-type. I crossed the two but the hybrid was sterile and when its meiosis was checked I realized that its parents differ by five chromosomal rearrangements. Accordingly, I described it as a new species, *Avena prostrata* (Ladizinsky 1971).

During my academic carrier I discovered a number of new species. The process by which they were discovered was similar in all these cases and was done in two stages: (1) detection of an unusual type which I have never seen before. (2) Hybridization experiments with what appears closely related species, to test if it is a new species or a variant of an already known species. This procedure complies with the biological species concept.

2.2.9.2 Avena damascena

The circumstances by which this species was collected were told to me by Tibor. Rajhathy, the one who actually collected this species. It was when a Canadian mission to collect *Avena* material visited Syria. While they drove about 60 km from Damascus in the Syrian Desert, an area with annual rainfall of 200–250 mm, he noticed in a wadi (dry water course) some plants which he believed to be *A. wiestii*. He started collecting seeds from these plants and immediately he found himself surrounded by Syrian soldiers who confiscated his belongings, but luckily he managed to pocket the seeds. He spent 2 days in custody and with the intervention of the Canadian Ambassador he was released but he had to leave Syria.

Back home he examined the karyotype of the collected seed. Most of them were diploids but some tetraploids. However, he noticed that the chromosome morphology of the diploid he examined was different from that of *A. wiestii*. He was unsuccessful in his attempts to cross it with *A. wiestii*, leading him to conclude that the diploid he had collected was a new species which he named *Avena damascena* (Rajhathy and Baum 1972).

I had no difficulty in crossing *A. damascena* with *A. strigosa*. The hybrid developed normally but its meiosis was irregular. The most frequent chromosome configuration in the pollen mother cells was a univalent, three bivalents, a trivalent, and a quadrivalent. These multivalents indicated that the two species diverge by two chromosomal rearrangements. However, in two out of the 100 cells examined two trivalents and a quadrivalent were observed indicating divergence of three chromosomal rearrangements between the two species. The hybrid was sterile but after its chromosomes were doubled by treatment with colchicine a fertile amphidiploids emerged. Chromosome association in meiosis of the artificially produced amphidiploids was regular with more than 12 bivalents and one quadrivalent per cell and it was fertile. The regular chromosome association is attributed to the preferential pairing potential of *A. strigosa* and it is likely that such a potential exists also in *A. damascena* (Cahana and Ladizinsky 1978).

For several years *A. damascena* was known from a single location in Syria. During a successful collection trip to Morocco, Mike Leggett and Per Hagberg collected seeds from several populations on the east-facing slopes of the Atlas range. Later, following hybridization with *A. damascena* from Syria, these proved to be of the same species (Leggett et al. 1992). This was a dramatic discovery because it indicated that *A. damascena* occurs at the two extremities of the Mediterranean basin. Furthermore, what actually it tells us about the origin and evolution of this oat species: did *A. damascena* originate spontaneously and independently in Syria and Morocco, and if not, where else might it occur?

2.2.9.3 Avena atlantica

The recently described taxon *A. atlantica* is also a member of series Eubarbatae. However, its dispersal unit is the entire spikelet (excluding the glumes). As already indicated, its taxonomic status should be regarded as a subspecies of *A. strigosa*.

I examined the *atlantica* type in its natural habitat in southwestern Morocco in an area with annual rainfall of about 200 mm. Even in a relatively rainy year it is rather rare and forms small and disjunct populations. The *atlantica* type was restricted mainly to sandy soil but some populations were found on brown soil.

How the spikelet disarticulation pattern was established in a member of series Eubarbatae is an interesting and intriguing question. Another question; what is the adaptive value of such a disarticulation pattern in the dry habitat where the *atlantica* type is a native? A possible reason for that could be survival improvement by prolonging germination ability. In oat species in which the dispersal unit is the whole spikelet, the seed of the lower floret germinates in the following winter while the seed of the upper florets stay dormant and might germinate in the second, third, or fourth year. Exceptions are the weedy forms of *A. sativa* in which all the seeds of the spikelet can germinate in the following year. I have not noticed this germination pattern in oat species in which the floret is the dispersal unit and I have always wondered how they cope with erratic rainfall common in the area where they are native. Further study is necessary for elucidating this question. Anyway, it has remained unclear why the spikelet disarticulation of ssp. *atlantica* occurred only in Southern Morocco and not in other relatively dry territories of the Mediterranean region where ssp. *wiestii* occurs.

2.3 Section Denticulatae

While the hexaploids are the most widespread oats of that Section, additional species, diploid, and tetraploids have been discovered in the last 40 years or so and are also characterized by dentate lemma tips.

The hexaploid oats manifest a wide range of variation in many morphological characteristics leading some botanists to suggest the arrangement of that diversity in terms of independent species.

Three types can be distinguished among the hexaploid oat. They are morphologically distinct from one another and also by their ecological requirements.

The first one is the *sterilis* type. It is characterized by a large dispersal unit consisting of the entire spikelet, with disarticulation occurring at the base of the lower floret only. The disarticulation scar is large with an oval-heart shape. It manifests great variation in size, lemma color, and hairiness. It is an aggressive weed and forms extensive populations in man-made habitats such as road sides, abandoned fields, and as weeds in cultivation. It also occurs in primary habitats, where no man intervention is evident, but such populations are much smaller.

The second is the *fatua* form in which disarticulation occurs at the base of each floret and the dispersal units are made of a single seed. This type is a prolific weed in cereal fields particularly in West Europe and North America. Rarely, it can be found in primary habitats, and if so, it is usually for a relatively short period. The *fatua* type also presents variation in lemma color and hairiness. Some forms are called Fatuoids and are believed to result from back mutation in the cultivated hexaploid oats.

The third type is the domesticated oats, the common oat. Here the variation is enormous because of human selection and breeding. Common oats do not shed their seeds but when the mature spikelet is forcibly separated into two florets, in some cultivars the rachilla remains attached to the upper floret (byzantina type), or to the lower floret, (sativa type). It is assumed that the byzantine type originated from the wild *sterilis* type and the sativa type from the wild *fatua* type.

Despite the enormous variability in the hexaploid oats they are all interfertile and should be regarded as one biological species *Avena sativa* with three sub pecies, *sativa*, *sterilis*, and *fatua*. This is a similar situation to the diploid *A. strigosa* and the tetraploid *A. barbata* which include wild weed and domesticated forms with wide range of diversity.

Classical taxonomic treatments of the hexaploid oats had proceeded in different directions, by separating each of the three types into several species. As happens in situations like that, different authors gave different names for their species. The treatment of the hexaploid oat in the monograph of Baum (1977) is a good example for that. For him, all the common oat cultivars have been regarded as members of a single species, *A. sativa*, while both the *sterilis* and the *fatua* types were divided to three species each. Even more astonishing, if not bizarre, he labeled all his hexaploid species as sharing the same genome, AACCDD.

2.3.1 The Designation of the Hexaploid Oats Genomes

Allopolyploids evolved by hybridization of species with lower chromosome numbers and chromosome doubling of the sterile interspecific hybrid. Detection of the progenitors of an allopolyploid species is usually achieved through an experimental procedure known as genome analysis (Lilienfeld and Kihara 1951). It is performed by crossing candidate species, diploids, or tetraploids to the allopolyploid species. If any of these candidate species appears to be cross-incompatible with the allopolyploid species, it will automatically be dropped from the list of the candidate species. In fact, I donot know any report that a diploid or tetraploid progenitor of an allopolyploid species was proved to be cross-incompatible with its allopolyploid derivative.

Once the suspected progenitor is found to be cross-compatible with the allopolyploid, the decision as to whether it is the progenitor or not is made according to the mode of chromosome association in meiosis of the interspecific hybrid. The progenitor is approved if its chromosomes pair regularly with those of the allopolyploid species. In addition, the number of chiasmata per cell in the interspecific hybrid should be similar or very close to that of the parental species with the lower chromosome number. Having the same number of chiasmata indicates that some chromosomes of the species with the higher chromosome number are sufficiently homologous to those of the candidate species to allow that pairing and chiasma formation. In this way, for example, the wild forms of *A. strigosa* were approved and accepted as the diploid progenitor of *A. barbata*.

The second step is to identify the individual chromosomes of the progenitor in the allopolyploid species. This is rather simple when the chromosomes of the two progenitors differ markedly in their size as in cotton. The cultivated cotton *Gossypium hirsutum* is allotetraploid, $2n = 4x = 52$, which originated from cross between African diploid cotton, *G. herbaceum* with relatively large chromosome, and *G. raimondii* from Peru with relatively small chromosomes.

When the chromosomes of the progenitors are not markedly different in their size or shape, accurate identification of the individual chromosomes of the progenitors in the allopolyploid species is laborious and requires special tester lines as had been done in wheat. Bread wheat *Triticum aestivum* is an allohexaploid, $2n = 6x = 42$ with the genomic constitution AABBDD. The A genome was contributed by the diploid wild wheat, *Triticum boeoticum* and the D genome by diploid *Triticum tauchii* (*Aegilops squrossa*). The identity of the third diploid which contributed the B genome has not yet been satisfactorily confirmed.

Throughout the years the distinguished wheat cytogeneticist E.R. Sears had developed complete sets of monosomic $(2n-1)$ lines of bread wheat in which one specific chromosome is presented in a single dose, and nullisomic $(2n-2)$ lines in which both homologous chromosomes are missing (Sears 1954). The 21 nulisomic lines were crossed with *T. boeoticum* to verify the chromosomes of the A genome in the hexaploid wheat. The chromosome number in all the 21 hybrids was 27 but with different pattern of chromosome pairing in meiosis. In hybrids involving 14 nullisomic lines, 7 bivalents, and 13 univalents were observed. This pairing pattern occurred in crosses with nullisomic lines in which the chromosome of B or D genomes were missing (Fig. 2.8). On the other hand, in hybrids of the other 7 nullisomic lines 6 bivalents and 15 univalents were observed. The additional univalents in that group indicated the absence of a specific A genome chromosome in the nullisomic lines (Fig. 2.8). The same procedure has been employed for

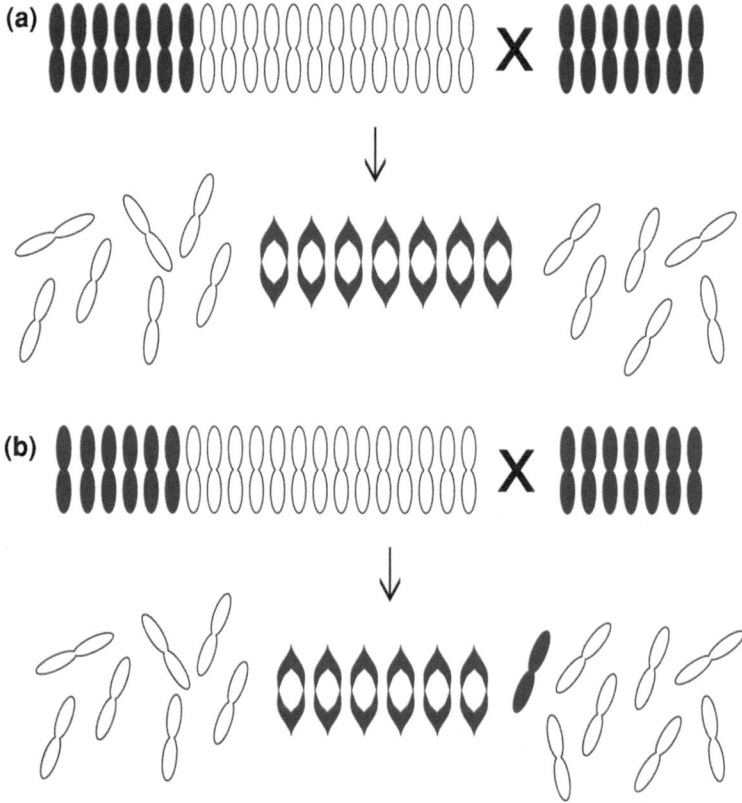

Fig. 2.8 Schematic drawing of metaphase I chromosomes in hybrids between wheat nulisomic lines and *Triticum boeoticum*, **a** the nulisone is of B or D genomes, **b** the nulisome is of A genome

detecting the D genome chromosomes by crossing 14 nullisomic lines, in which individual chromosomes of D and B genomes are missing, with *Ae. squarossa*. After identifying the chromosomes of the D genome, the rest have been considered by subtraction as being B genome chromosomes.

The attempt to identify the chromosomes of the hexaploid oat genomes has proceeded in a way that is both peculiar and wrong. Nishiyama (1929) was the first to suggest that the hexaploid oats is made of three genomes, AB and C, and those of *A. barbata* being AB. By comparing chromosome morphology of the hexaploid oats and of *A. barbata*, Rajhathy and Morisson (1959) showed that the hexaploid oats do not possess the chromosomes of the B genome of *A. barbata* and designated the genomes of the hexaploid oats as ACD. Of the 42 *A. sativa* chromosomes they selected 14 with the morphologically appearance of *A. strigosa* chromosomes and labeled them as the A genome. The other 28 chromosomes were lumped together as C and D genomes. These authors presented no evidence that the selected 14 chromosomes are genetically homologous to *A. strigosa* chromosomes, or at least, comprise a coherent genome. They did not cross the *A. sativa* line that they used with *A. strigosa* because the two are cross incompatible and successful hybridization between them requires transferral of the young embryos to culture medium. Therefore, they crossed the *A. sativa* line with artificial autotetraploid *A. strigosa* and *A. barbata*. They found many cells with 13–14 univalents but admitted that the pairing pattern could be a result of autosyndesis (pairing between chromosomes of the same species) and therefore inconclusive (Rajhathy and Morrison 1960).

On the other hand, when hybrids between the common oat and *A. strigosa* have been produced, by the aid of embryo culture, they manifested poor pairing, 2–5 bivalents, almost all with a single chiasma, and 1–3 trivalents per cell. As already indicated, *A. strigosa* chromosomes have the tendency toward preferential pairing in the presence of homologous chromosomes. If the *A. strigosa* genome is indeed presented in the common oat one would expect similar pairing, yielding seven bivalents, in the tetraploid hybrid as in the diploid species. This did not happen and furthermore, the number of chiasmata in the hybrid was only half their number in *A. strigosa* (Aung et al. 2010).

By similar technique of comparing chromosome morphology, Rajhathy and Sadasivaiha (1968) selected 14 chromosomes of the tetraploid *A. magna* and labeled them as A genome and another 14 as being of C genome supposedly donated by *A. clauda*. I myself had been lead astray by Rajhathy suggestion and when I examined the karyotype of *A. ventricosa* and realized that its chromosomes are acrocentric, I assumed that the C genome was contributed by *A. ventrticosa*. Only later, when it became obvious to me that the identity of the hexaploid oat genomes cannot be resolved by karyomorphological comparisons, I withdrew my previous suggestion.

Using the embryo culture technique, Leggett (1998) crossed *A. strigosa* with *A. clauda* ssp. *eriantha*. Chromosome pairing completely failed in 17.65 % of the cells in the hybrids and not more than 4 bivalents were recorded in about 76 % of the cells. By colchicine treatment a $2n = 27$ (monosomic) plant was obtained in which chromosome pairing was nearly regular, 13 bivalents and a univalent were recorded in over 45 % of the cells. This is not surprising because of the marked tendency of *A. strigosa* chromosomes to pair with their homologs in the presence of chromosomes of another genomes which are only partially homologous to *A. strigosa* chromosomes (2.3.8). In addition, lemma tips in the artificial allotetraploid terminated in two bristles and were not denticulate as in *A. magna* (Leggett pers. Comm.) When this artificial allotetraploid was crossed with *A. magna* (supposedly an AC genome) chromosome pairing was highly irregular. Maximum pairing was a quadrivalent, two trivalents, six bivalents, and five univalents, and the minimum, 4 bivalents and 19 univalents. This paring pattern clearly indicates that the two *A. magna* genomes are considerably differentiated from *A. strigosa* A genome and *A. clauda* C genome and could not originate from these two diploids.

Another attempt to identify the three hexaploid oats genomes was made by genomic DNA in situ hybridization (Jellen et al. 1994, Chen and Armstrong 1994). By this technique, however, *A. strigosa* genome (AA) could not be separated from modified A genomes of *A. canariensis, A. damascena A. longiglumis*, and *A.pro-strata*. Similarly, *A. clauda* and *A. ventricosa* commonly referred to CC genome were indistinguishable from one another. When the hexaploid oat chromosomes were hybridized with DNA of A genome species, original or modified, 28 of the hexaploid oat chromosomes showed complete hybridization. Another 14 chromosomes showed hybridization with genomic DNA of *A. clauda*. Accordingly, it has been proposed that A and D genomes of the hexaploid oats are identical and probably evolved through autopolyploidy (Jellen et al. 1994; Leggett and Markhand 1997).

The two approaches for estimating genome relatedness, chromosome pairing in interspecific hybrids and genome in situ hybridization, are therefore pointing to different directions. Obviously the latter is less sensitive because its inability to separate the various A genome types, which has been accomplished successfully by the former approach. Hybridization experiments involving the diploid oats species and the hexaploid and tetraploid oat species and analysis of the chromosome pairing in these hybrids do not support any of the species with the alleged A or C genomes as participants in the origin of these polyploid oats. If nevertheless, those genomes did participate in the origin of the hexaploid oats why does the cytogenetic evidence not show it, and why does that kind of evidence corroborate the involvement of *A. strigosa* in the origin of *A. barbata*.

By now it is obvious that the ACD genomic formula of the hexaploid oats as it perceived by many oat specialists is totally unfounded and the alleged diploid progenitors with A and C genomes did not participated in the origin of the tetraploid species whose genomes are labeled as AC, and nor in the origin of the hexaploid oats. It is unfortunate that this erroneous and misleading genome labeling is being copied from one article to another and by so many authors. This not only helps the establishment of false ideas, but it also serves to misdirect further research. It is the time to acknowledge the fact that none of the presently known diploid oat species had participated in the origin of the polyploid species of section Denticulatae. Until the true genome donors to these polyploids are discovered, all that is said or written about their genomic formulas will remain speculative and unfounded.

Morphological and ecological evidence also indicate that *A. strigosa* and *A. clauda* could not have participated in the formation of the tetraploid oats *A. magna, A.murphyi*, and *A. insularis* which are erroneously regarded as having AC genomes. The dispersal units of the latter three species contain the entire spikelet (glumes excluded), are large and coarse with big disarticulation scars that are rather blunt at the bottom. On the other hand, in the two alleged diploid progenitors, wild *A. strigosa* and *A. clauda*, the dispersal units are small and the disarticulation scar is with rather sharp bottom. In addition, lemma tips in the three tetraploids are dentate while in the diploids they are aristulate.

What then is the origin of the unique spikelet characteristics of these three tetraploids and also of the hexaploid ssp. *sterilis*? The unique spikelet morphology of the three tetraploids is obviously not designed to facilitate taxonomy, but is of a clear adaptive value. These tetraploids are adapted and restricted to heavy crumbling, black–brown soil which cracks when dry. That adaptation is so specific that in nature they cannot grow in other soil types. At maturity, when the plants shed their seeds, the large dispersal units easily and quickly become buried in the ground. Even if they fall into soil cracks their relatively large caryopsis enables the seedling to emerge from depth of 10–15 cm and even more. In that kind of soil *A. clauda* and the wild forms of *A. strigosa* have never been observed. Perhaps, the reason for this is that their dispersal units may burry themselves too deep to be able to emerge the following year. In fact, none of the presently known diploid oat species is adapted to the soil type in the natural habitat of *A. magna, A. murphy*i and *A. insularis*. While the hexaploid ssp. *sterilis* is also adapted to that kind of soil, it is ecologically much more flexible and grows successfully in other soil types even with hard surface where the large dispersal units cannot burry themselves in the ground but can conceal themselves under stones or organic residues. Such protection is by no means perfect and a large proportion of the dispersal units are eaten by ants and rodents. Nevertheless, this is apparently not a critical problem for ssp. *sterilis* which usually grows in massive stands and produces a large number of seeds.

2.3.2 The Discovery of Additional Species in Section Denticulatae

2.3.2.1 Avena canariensis

This is the only diploid oat species with denticulate lemma tips. Initially it was detected following examination of herbarium material by B.Baum. When it was later discovered in the Canary Islands Fuerteventura and Lanzarote, many believed that it is the diploid progenitor of the hexaploid oats. This hypothesis was disproved according to the chromosome pairing in the *A. canariensis* x *A. sativa* hybrid which was even inferior to *A. strigosa* x *A. sativa* hybrid (Thomas 1992).

2.3.2.2 Avena agadiriana

Like *A. canariensis*, this species was discovered following a survey of herbarium material. It was found in littoral southwest Morocco, mainly on sandy soil and also on stony brown soil. As it might seem odd, cytogenetically it is closer to *A. barbata* than to *A. magna*, particularly by the significantly smaller number of univalents in the hybrids involving the latter (Leggett personal communication). While some oat specialists have suggested that the *A. a*g*adiriana* genomes are similar to those of *A. barbata*, perhaps because of in situ hybridization experiments, cytogenetic and morphological evidence are not in favor of such an idea. More research is needed to clarify the nature and the evolutionary relations between this species and other diploid and tetraploid oats.

2.3.2.3 Avena magna

In 1964, three oat specialists, F. J. Zillinsky from USA, T. Rajhathy from Canada, and A. Hayes from UK traveled to Morocco and Algeria to collect wild oat species. The collected material was divided between the parties and each of them conducted different studies on the material. In the USA, the *sterilis* material was examined for two main characteristics, crown rust resistance and protein content of the groat. One specific accession was outstanding with respect to both characteristics and was crossed with a common oat cultivar. Surprisingly, the hybrid was sterile. When its chromosome number was checked it appeared to be pentaploid ($2n = 35$) and its wild parent to be tetraploid. In 1968 they announced it as a new tetraploid oat, *A. magna* (Murphy et al. 1968). For the oat workers of that time the discovery of a *sterilis*-like tetraploid oat was sensational news. For my part, I was extremely pleased with that discovery because more or less at that time, I had predicted that a *sterilis*-like tetraploid oat must exist. This was because the results of a comparative morphological study that I conducted in the genus made me realize that the morphological characteristics of ssp. *sterilis*, the main taxon of the hexaploid oats, were missing in any of the oat species which were known at that time. I also speculated about the habitat of that as yet hypothetical species and assumed that it grew on heavy soil. In their paper of 1968 there was no mention of the habitat where *A. magna* was collected, only the location, between Oulmes and Tiflet, approximately 30 km southeast of Tiflet, and the name of the collector, F.J.Zillinsky.

In an attempt to get information on the habitat from where *A. magna* was collected I contacted Tibor Rajhathy with whom I used to correspond from time to time. He informed me that on the day that *A. magna* was collected each of them had taken a different route and the one who had actually collected the material was A. Hayes. I wrote to A. Hayes asking about the soil type where he collected *A. magna*. He replied saying that he cannot remember but he assumed that it was heavy soil. This was a remark that I waiting for. He also mentioned that he stopped at that particular site because there was a rose plot next to the road and after looking at the roses he collected a few spikelets from the ground along the road ditch. He also enclosed a photograph of himself near the roses and a vast plain in the background.

Several months after the *A. magna* publication, a letter to the editor was published in Science on this matter by three authors. One of them, G. Martinoli (1969), mentioned that he discovered *A. magna* already in 1955 in Sardinia (Martinoli 1955). In fact, he referred to a *sterilis*-like tetraploid he had collected at three sites in Sardinia near Teulada, Serramanna, and Geremeas. These were all the *A. sterilis* plants that he had examined cytologically and according to him all were tetraploids. The photograph of the chromosome in that paper, apparently the best he had, was of poor quality and he added a hand drawing of the chromosomes of the so-called tetraploid *A. sterilis*.

This new information thrilled me. If the three *sterilis* specimens that he had examined cytologically were collected in different parts of the Island and all were tetraploids, this type apparently occurred all over the island. Instantly, I decided to go to Sardinia and check this type in its natural habitat. I took with me a microscope and all the equipment necessary for counting chromosomes. I arrived in the spring when the oat plants were at the beginning of flowering and individual plants had developed panicles and tillers at flag leaf stage (suitable for checking meiosis). My disappointment began when I saw the first wild oats there. They looked to me like the ordinary hexaploid which I had verified by chromosome counts. The same was near Teulada, Serramanna, and Geremeas where Martinoli had claimed that the tetraploid *A. sterilis* was collected.

In Sardinia I came across other wild oat species, *A. barbata* and *A. fatua* which Martinoli had mentioned in his paper, with correct chromosome numbers, but not tetraploid *A. sterilis*.

2.3.2.4 The Origin of the Epithet *Avena maroccana*

In his monograph, Baum (1977) treated *A. magna* as a heterotypic synonym of *A. maroccana* Gdge. It seemed odd to me and I wrote a paper on that (Ladizinsky 1993) but in my mind I do not think it got the publicity it deserved.

In his examination of many oat herbarium specimens Baum had come across a number of odd types some of which such as *A. canariensis* and *A. agadiriana* later became species on valid grounds. One of these odd types was *Avena maroccana*. That specimen had been collected by a French botanists name Gandoger. It seems that this botanist visited littoral Morocco twice. In 1903 he visited Ceuta. At that time, as he noted (Gandoger 1907), there was a period of unrest in the region and strangers were not allowed to go outside the town. Nevertheless, escorted by Spanish soldiers, he managed to collect some plants in the vicinity of the town. Among the collected plants he listed three wild oat species, *A. fatua*, *A. sterilis*, and *A. longiglumis*. The latter species as already indicated occurs along the coast of the Mediterranean Sea, but exclusively on sandy and sandy loam soils. In 1908 Gandoger made another trip to littoral Morocco, this time visiting the Chafarinas Islands near Melilla from where he mentioned *A. maroccana* and provided a description of that new taxon. Strangely enough, he referred to Ceuta as the type locality "sterilibus herbosis maritimis circa Ceuta", and the date of collection was given as 25 April 1903 (Gandoger 1908). On technical grounds I am not sure that *A. maroccana* is a legitimate taxon, but this is a minor issue.

The main point is that in Ceuta, Gandoger was under pressure from the soldiers who escorted him to return to the safety of the city walls as soon as possible, as he vividly described in his paper, and could not get too far from Ceuta. In the area which he managed to explore he collected *A. longiglumis*, indicating that it was sandy or sandy loam soil, where ssp. *sterile* and ssp. *fatua*can also grow, but not *A. magna*. Even at the distance of a day's walk from Ceuta and beyond, the habitat of *A. magna* is not found.

The morphology of *A. maroccana* is also disturbing. A photograph of *A. maroccana* herbarium sheet is presented in Baum's monograph and the spikelets

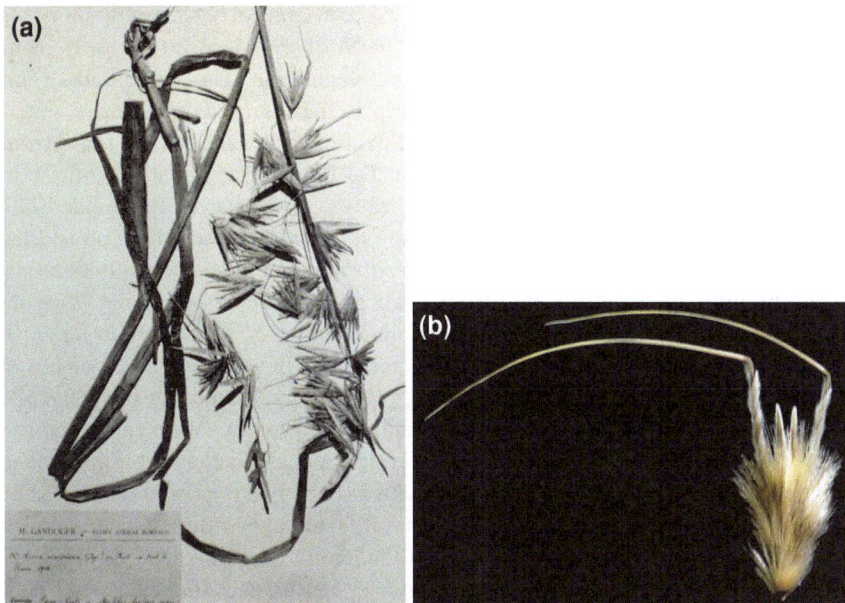

Fig. 2.9 **a** The type specimen of *A. maroccana*, **b** spikelet of *A. magna*

are of typical V shape as in ssp. *sterilis*, whereas in *A. magna* the lemma tips of the
two lower florets are close and nearly parallel to one another. At that point it
seemed necessary for me to examine personally the type specimen of *A. maroccana*. Baum (1977) had mentioned that the holotype is deposited in the herbarium
of Lyon, France. I contacted the keeper of the herbarium in Lyon on this matter,
but he informed me that this particular specimen had disappeared after being
inspected by Baum. On the other hand, Baum told me that he did not take that
specimen on loan. So, all that is left is the photograph of the type specimen which
is looked like ssp. *sterilis* and not *A. magna* (Fig. 2.9).

The third point is the chromosome number. This was not mentioned by
Gandoger, and obviously not by Baum. After all, the most exciting part of the
discovery of *A. magna* was the $2n = 28$ chromosome number which is of marked
biological and diagnostic significance.

I can only hope that now, after the story of *A. maroccana* has been told oat
workers will no longer use this epithet and the credit will revert to those who really
discovered *A. magna*.

2.3.2.5 Avena murphyi

In 1968, when the discovery of *A. magna* was published I could not go to Morocco
for a field study of this new species. No Israeli citizen could get a visa to Morocco

at that time. It seemed to me that searching for *A. magna* in Southern Spain might be fruitful because this territory is similar to Northern Morocco not only in its climate but also in its geology, soil type, and vegetation. Therefore, finding this species was another aim of my collection trip to the Iberian Peninsula.

I arrived in Tarifa at the southern tip of Spain in early June but the season was over for oats. When I drove on the road from Tarifa to Cadiz, about 12 km NW of Tarifa, I reached a vast valley with black heavy soil. The area was divided into fenced plots and was used for grazing. Along the road, between the road and the fence, there was a massive stands of ssp. *sterilis* that had already shed their seeds. Inside the plots there were a lot of perennial grasses such as *Phalaris tuberosa, Hordeum bulbosum, Lolium perenne*, and others, together with *Hedysarum coronarium* a legume with an attractive red inflorescence. In addition, some patches of oat plants were scattered inside the grazing plots. The soil type and the landscape in that valley reminded me of the photograph sent to me by A. Hayes of the site where he collected *A. magna*. I stopped to collect some spikelets from the ground and the first one I picked up was peculiar, with a shape I had not seen before. It had a large disarticulation scar, glabrous lemmas, with awn inserted at the lower quarter of the lemma, and the upper part of the lemmas tending toward each other and then backward, forming rather a ventricose appearance. Immediately, I realized that this was not *A. magna* but at that time I did not know what it was. I looked for more spikelets of the same type and collected about 12 of them from an area of about one square meter which indicated to me that they all probably originated from a single plant. I stopped several more times in the valley but did not find that oat type again.

Back home, I germinated two seeds of this strange type in order to count chromosomes in root tips. Surprisingly, it turned out to be tetraploid. When these plants started flowering they were crossed with *A. magna* and *A. sativa*. The hybrids with both species had irregular meiosis and were sterile. This indicated to me that the strange oat type was well differentiated from *A. magna* and deserved a status of a new species and also that it did not participate in the formation of the hexaploid oats.

These conclusions were based, however, on behavior of a single plant and I felt that it was important to obtain more information about this type. So, I went back to the Tarifa area at the southern tip of Spain, this time at the beginning of May. I found this oat type growing everywhere, along the road sides and inside the grazing plots. Most of the plants had glabrous lemmas but some had hairy lemmas. Along the road sides it grew together with ssp. *sterilis*, but in the grazing plots in pure stands. While it was rather common in the valley, I could not find it in other places in the same area.

I prepared a manuscript on the discovery of the new oat species which I called *Avena murphyi*, after H.C. Murphy who had arranged the fund for my study. He had tragically drowned while he was on vacation. I submitted it to *Science* because the *A. magna* story was published there. To my surprise it was rejected by two reviewers. One claimed that I did not follow the style that was used for *A. magna*, and the other came with the bizarre reason that botanically Spain is known so well that it was impossible that new species would be found there. The manuscript was submitted to another journal. This time it was sent for review to a grass specialist in Kew Garden. He wrote back, publish it immediately because it is sensational.

2.3.2.6 Avena insularis

In 1994 I read an article on Sicily in the *National Geographic* Magazine. In one of the pages there was a photograph of *Hedysarum coronarium*,(suja),which instantly reminded me the association between this legume and *A. murphyi* in Southern Spain. I asked the writer of that article where she had photographed the suja flower. She replied, this was the national flower of Sicily, and could be found everywhere. I came to Sicily in order to try to find *A. murphyi* there. I hoped to find it on uncultivated heavy soil, but on the first day of my excursion from Catania to Gela all the land with heavy soil that I could see was under cultivation.

Next day I continued toward Agrigento, and shortly after leaving the city limit of Gela I noticed in a distance of about 3 km, hills of heavy soil and because of the rough terrain they had not been cultivated. There was no direct road leading to these hills and it took me some time to get close to them. I had to walk several hundred meters in the plain before starting to climb the hills. There were numerous ssp. *sterilis* plants on the plain, mainly in cultivated plots or adjacent to them, but not on the hills. Then I noticed several groups of oat plants scattered on the slopes, but they were shorter than ssp. *sterilis*, had smaller panicles, and each spikelet connected by its own pedicel to the panicle axis. When I looked at the spikelets they were also different. They were smaller compared to those of *A. sterilis* in that area, and although they were V-shaped, it was not as pronounced as in *A. sterilis*. In addition, the disarticulation scar was longer and narrower than in ssp. *sterilis*. I collected some spikelets for further examination and showed them to my wife who was waiting for me in the car. She looked at them and said why did you have to go so far? This type grows right here. She was right, several plants of this type were growing on a slope next to the car.

I wondered if this type had ever been collected before. I returned to Catania and looked at the oat collection in the herbarium of the Botany Department at the University of Catania. There were over 100 herbarium sheets of oats there, most of them of ssp. *sterilis*. None of them was similar to the type which I had found near Gela. I wanted to check the oat collection at the University of Palermo but they provided me only ten herbarium sheets and when I mentioned that in Catania I examine about 100 sheets they added another ten. They all were typical ssp. *sterilis* plants.

As soon as I could I germinated the seeds brought back from Sicily and checked their chromosome number. It was 28. When the plants reached flowering they were crossed with *A. magna* and *A. sativa*.

The hybrids with *A. magna* had irregular chromosome pairing at meiosis and were sterile (Ladizinsky 1998), indicating that this new tetraploid is well differentiated from *A. magna*. However, the most surprising findings were the hybrids with *A.sativa*. The pentaploid ($2n = 35$) hybrids were obtained only when *A. sativa* was the female parent. This also has been the case in hybrids between the common oat and both *A. magna* and *A. murphyi*. Two hybrid combinations were tested involving different common oat cultivars. In one combination the chromosome configuration in metaphase 1 of meiosis consisted of 4.8 univalents (I), 10.25 bivalents (II), 1.92 trivalents (III), 0.86 quadrivalents (IV), and 0.09 higher multivalents. In the other combination the average chromosome configuration was 5.87 I, 9.87 II, 1.29 III, 0.16 IV and 0.16 higher multivalent. The average number of chiasmata per cell was 22.40 (19–26) in the first combination and 23.16 (21–26) in the second combination. Pollen fertility was above 30 % in these hybrids and some seeds were set by selfing.

The data of chromosome pairing and fertility of these hybrids were truly astonishing because they were much better than in the hybrids between the hexaploid oats and both *A. magna* and *A. murphyi*. Instantly I realized that the tetraploid oat from Sicily was probably the progenitor of hexaploid oats and for three reasons: First, in the new tetraploid an average number of chiasmata per cell was 26.13. In the pentaploid hybrid involving this tetraploid and the common oat, the average number of chiasmata per cell was 23.16. In other words, 88 % of the pairing potential of the new tetraploid has been materialized in the pentaploid. For comparison, in pentaploid hybrids involving *A. magna* 75 % of the paring potential has been expressed and in hybrids involving *A. murphyi* and *A. barbata* the respective pairing potential were only 62 and 42 % (Ladizinsky 1998). Obviously, the new tetraploid is more likely than any of the other three tetraploid species to be the progenitor of the hexaploid oat.

Second, the partial fertility of the pentaploid hybrids involving the new tetraploid was also unique. Pentaploid hybrids involving the other three tetraploid species had no normal pollen grains and could produce a small number of seeds only by massive pollination with the parental species pollen. In contrast, the pentaploid hybrid involving the new tetraploid had over 30 % of regularly shaped pollen grains which were as intensively stained by acetocarmine as its parental pollen grains. Moreover, it produced seeds by selfing. Similar fertility has been observed in the triploid hybrids between *A. barbata* and *A. strigosa* which is the diploid progenitor of the former species. The analogy of chromosome pairing and fertility in both cases further support the new tetraploid as the progenitor of the hexaploid oats.

Third, mature spikelets of *A. sterilis*, the most widespread hexaploid oat is typically V-shaped. Both *A. magna* and *A. murphyi* were different in this respect, but mature spikelets of the new tetraploid are also V-shaped, though not as pronounced as in ssp. *sterilis*

While the accumulated evidence indicated that the tetraploid oat from Sicily was a new species and the progenitor of the hexaploid oats, the examined plants were derived from one small population. More information about its ecology and distribution in Sicily was essential before it could be declared a new species.

On a subsequent visit to Sicily I explored the area around Lake Comunelli, NW of Gela, where uncultivated heavy clay soil was common. Because of the

Fig. 2.10 The habitat of *A. insularis*

inhospitable terrain, the hills were not cultivated and it seemed that some of the fields had not been in use for several years. The soil was overlaying colorful sand stone and occasionally that color could be traced in the clay soil. In some places the soil was badly cracked, probably having remained so from the previous summer, and I had to watch my step to avoid falling into them. The vegetation was almost purely herbaceous with no trees. The dominant perennial grass was *Lygeum spartum* (Fig. 2.10) which normally occupies dry habitats. In some sites the new tetraploid oat was the most common annual grass and formed nearly pure stands. Occasionally, suja plants of, *Hedysarum coronarium*, were found growing among the oat plants. The occurrence suja plants and *Lygeum spartum* indicates the extreme and variable water regime in the habitat of the new tetraploid. The first companion plant was confined to wet habitats while the second one to dry habitats. The Forestry Department of Sicily had planted eucalyptus trees in some places where the new tetraploid was growing. As a result, ssp. *sterilis* had invaded the habitat and mixed stands of the two oats were not uncommon. In sites such as these it was not always possible to distinguish between the two, again, indicating gene flow as expected from the partial fertility of the hybrids between them.

Near Lake Comunelli I detected four major populations of the new tetraploid, in addition to several smaller populations. In another area, about 16 km NE of Lake Comunelli, near Mt. Bubonia, an additional population was found at the edge of eucalyptus plantation.

Hybrids between representatives of Lake Comunelli populations were normal and fertile but hybrids between them and of Mt. Bubonia were partially sterile but no meiotic irregularities were observed.

By this time it became clear to me that the new tetraploid was an important new species, and probably the progenitor of the hexaploid oats, and was named *Avena insularis*. So, I had gone to Sicily to look for *A. murphyi* and discovered the tetraploid progenitor of the hexaploid oats, a species whose existence I had predicted 30 years earlier. For a moment I felt like Saul who went on a search for his father's lost asses and found the monarchy.

Although it was named *A. insularis* (from the Island) it seemed to me unlikely that it had originated in Sicily. The main reason was that no diploid oat species had been reported from the island. Thus, if it had indeed been brought to Sicily, questions remain about when and how it had come from.

Exploration and collection of wild oat species in North Africa during the last 50 years had pointed to that area as the main center of diversity of the genus *Avena*. Therefore, it was not unreasonable to assume that *A. insularis* had arrived in Sicily from that region, most likely from Tunisia because of the short distance between them.

Before going to Tunisia I studied the geology, soil types, and vegetation of the region. By the aid of detailed soil map I noticed a few locations that looked promising for a search of populations of *A. insularis*. The first one was near Temime, at the horn of Tunisia. In that area I came across a valley of heavy clay soil. The vegetation was lush because of abundant rainfall. Subspecies *sterilis* was everywhere but there was no sign of *A. insularis*. The second area was near Bargou, formerly Robaa, NE of Siliana about 150 km SW of Temime area. The soil was similar and overlaying sand stone. The whole area was extensively cultivated with wheat as the main crop. I stopped near an uncultivated patch between the road and wheat field which by then was protected from grazing. Plants of *ssp. sterilis* were common in the wheat field but not in the uncultivated patch. I noticed suja plants, which were in an early stage of flowering, and immediately after that I came across a number of oat plants similar to *A. insularis* but smaller and more delicate. Their seeds, unfortunately, were immature. I continued on my way and explored other areas with clay soil but on calcareous bed rock could not find any *A. insularis* plants. Before leaving Tunisia I returned to Bargou for seed collection but they were still immature. I uprooted a number of plants, put them in a plastic bag and planted in the green house soon after arriving home. One weekend there was a power cut and the temperature shot up, over 50°, and all the plants which were brought from Tunisia died.

The following year I returned to Tunisia and went straight to Bargou. This time it was a dry year, the land was bare with a few and small annual plants scattered over the site. The few *A. insularis* plants I found in the uncultivated site were not taller than 25 cm, but at least they had mature seeds. Around Bargou I found more populations, always in the same soil type and in uncultivated land, sometimes along narrow strips between wheat fields and dry water course, where they were protected from grazing. It appears that grazing represents the most serious threat to *A. insularis* in Tunisia.

After thoroughly searching the Bargou area I drove back to Temime to explore the hills around the valley. The soil on the hills was practically bare but small and delicate plants of *A. insularis* were found growing individually or in small groups.

Now that I had some *A. insularis* seeds from Sicily and Tunisia, the time had come to try to answer the questions about the origin of the Sicilian populations. Once again hybridization experiments were the tool (Ladizinsky and Jellen 2003). The Bargou and Temime populations appeared to diverge from one another by a major chromosome translocation as indicated by the formation of a quadrivalent or trivalent and univalent at metaphase I of meiosis. Such a chromosomal aberration usually causes 50 % fertility, but in this case pollen fertility ranged from 20 to 30 % and seed set from 5 to 8 %. Thus, it seemed that other genetic factors were playing a role in the low fertility.

Hybrids of Temime x Gela cross had normal meiosis and about 90 % stained pollen grains, though 20–30 % were smaller than the rest. Seed set was low, 27–55 % among different plants. Surprisingly, no seeds were produced in the reciprocal cross.

Chromosome association in Bargou x Gela hybrids was 12 bivalents and a quadrivalent, indicating that the chromosomal translocation occurred in the Bargou population of *A. insularis*. Nevertheless, seed set in this hybrid combination was higher, 21 %, compared to Temime x bargou hybrids. It thus appears that in each territory there was genetic sterility between populations, and that there was chromosomal sterility between the Tunisian populations. This sterility was even more pronounced between the Sicilian and the Tunisian populations. Accordingly, it is not likely that the Sicilian populations of *A. insularis* originated directly from the Tunisian populations that have been examined. It is possible, however, that they originated from other populations in Tunisia that have not yet been discovered.

Even if the Sicilian population had originated from Tunisia, the way they arrived in Sicily is not clear. Sicily has never been connected to Africa by a land strip. In fact, the depth of the sea there is over 2000 m. Flotation by sea water is unlikely because the Temime populations grow about 10 km from the shore of the Mediterranean Sea. Transport by migrating birds can also be ruled out because bird's migration from Africa to Europe occurs at early spring when the oats are just flowering.

Another possibility is transport by humans but this is also problematic. Crossing of the Mediterranean Sea by boats started at the early history. By the sixth century BC the Carthaginians established their main center in Tunisia and controlled part of Sicily. After the Punic war between Rome and Carthage and the destruction of the Carthaginian kingdom, Tunisia became part of the Roman Empire and its agriculture resources were exploited for the benefit of Rome. Most likely, cereal grains had been an important commodity exported from Tunisia to Sicily. If so, could seeds of *A. insularis* have been brought to Sicily as a contaminant in wheat and barley? Such a scenario is, however, unlikely since this wild oat is not a weed and does not exist on arable land. A more reasonable scenario for the arrival of this wild oat from Tunisia to Sicily, if it had happened at all, is that it was cut for fodder for livestock during transportation. Nevertheless, the short time, in evolutionary terms that has elapsed from the hypothetical introduction during Roman times was probably insufficient for the establishment of the genetic differences that reduce the fertility of hybrids between the populations of these territories. So, unfortunately, we still do not know when and how *A. insularis* arrived in Sicily.

2.4 Back to Sardinia

After the discovery of *A. murphyi* in Spain and *A. insularis* in Sicily I was wondering, after all, was Martinoli correct in claiming that he collected and examined a tetraploid sterilis-like type in Sardinia. I return to Sardinia to check it again, this time with a much better knowledge of the ecological preferences of the sterilis-like tetraploid oat species. In 1997, I revisited the three areas where according to Martinoli the $2n = 28$ sterilis-like oat had been found. The ecological differences between Gela in Sicily and the three sites in Sardinia were striking. The bedrock is granite in the Geremeas area, but metamorphic in Teulada. The soil at both sites is shallow and compact and the vegetation is typical Mediterranean with a lot of *Pistacia lentiscus, Calicotome villosa* and *Cistus* sp. Serramanna is located in a vast cultivated plain. The soil is deeper than at the other two sites but equally compact. In all three sites ssp. *sterilis* plants were confined to road sides and abandoned cultivations. In recently fallowed fields the ssp. *sterilis* plants were well developed and grew in large stands. Where cultivated fields had been left for greater length of time, only a few ssp. *sterilis* plants were found. They had short stems and panicles with more slender spikelets, and fewer spikelets per panicle. Representative chromosome counts in these plants showed them all to be hexaploids. It seems, therefore, that the confusion with *A. barbata*, as suggested by Rajhathy (1969), remains the most plausible explanation for Martinoli's "tetraploid *A. sterilis*".

2.5 Cytogenetic Affinities Between *A. magna,* *A. murpyhi, A. insularis* and Their Possible Evolution

Chromosome pairing in hybrids involving these three tetraploid species indicates complex relationships among them. There is a considerable chromosome repatterning and reduction of homology between any two species as indicated by the occurrence of 3–5 trivalents, 1–3 quadrivalents, and 1–2 higher multivalent per cell, in addition to 0–11 univalents (Table 2.3). Such cytogenetic relationships imply that they could not have evolved from one another or from a common ancestor simply via successive accumulation of a small number of chromosomal aberrations or reduction of homology between specific chromosomes. A more likely Hypothesis is that they emerged directly and spontaneously from a common ancestor, or that one of them gave rise to the other two. However, such a scenario

Table 2.3 Summary of cytogenetic relationships among the tetraploid wild oats *A. magna, A. murphyi* and *A. insularis*. I univalent, II bivalent, III trivalent, IV quadrivalent, V pentavalent

Cross combination	I	II	III	IV	V	Xta/cell
A. insularis x *A. magna*	2.19	7.73	1.36	1.25	0.15	19.28
A. insularis x *A. murphyi*	4.80	9.33	1.05	0.35	0.10	18.8
A.magna x *A. murphyi*	6.23	5.37	1.73	1.14	0.21	16.38

requires a major genome breakdown and regrouping of the chromosomes in a new way. Although this might sound unlikely, there are a number of reports of spontaneous massive chromosome breakage in individual plants such as in *Elymus fractus* (Heneen 1963) and *Aegilops longissima* (Feldman and Strauss 1983). Some individual plants of an *Ae. longissima* population from southern Judea hills, Israel, have shown extraordinary breakage of mitotic and meiotic chromosomes. This affected all the chromosomes but in a non-random fashion. The number and the frequency of the observed aberrations were increased by elevating the temperature from 25 to 30–32°C. This chromosome breakage seemed heritable and was observed in the few progenies produced by these plants. Furthermore, in crosses with normal plants chromosome breakage appeared to be governed by a single recessive gene, and the proportion of the chromosome breakage by the homozygous gene was affected by the cross direction, pointing to cytoplasmic involvement. Another important aspect of the massive chromosomal rearrangements in both *Elymus* and *Aegilops* is that they have not entailed any obvious morphological changes.

Massive chromosomal rearrangement between closely related species may thus indicate a major genome rearrangement and the examples are not uncommon also in the genus *Avena*. As already mentioned, the cytogenetic affinities between *A. strigosa* and *A. prostrata* are a clear example of such an event. These two species differ by five chromosomal rearrangements, yet their morphology is about similar. If these rearrangements were gradually accumulated, one would expect to find four intermediate populations, each with additional rearrangement. So far this has not been recorded. Nevertheless, it is unlikely that in such scenario the original and the current populations are still in existence and prosperous, while all the intermediate links have disappeared. *A. strigosa* differs from *A. longiglumis* by a similar number of rearrangements but in this case the morphological change is noticeable. On the other hand, the soil preferences of the latter are within the range of habitats occupied by wild *A. strigosa*.

A number of characters of *A. magna, A. murphyi*, and *A. insularis* favor the hypothesis of genome breakdown and chromosome regrouping during their evolution. They are all adapted to the same soil type and are nonexistent in other soil types. They share the same gross morphology with small differences between them. Geographically they are found in a relatively small territory and occasionally grow in mixed stands, at least the first two species in Tangier area of Morocco. In situ hybridization of *A. magna* chromosomes with DNA of *A. strigosa* and *A. clauda* showed five translocations between the two *A. magna* genomes Leggett

et al. (1994). Such translocations have not been reported in hybrids between
A. magna accessions and they may indicate genome breakdown in the formation of
the three tetraploid species.

Another possible reason for the complex cytogenetic relationships between the
three tetraploids could be that each of them evolved from different diploid species.
If this is the case, one would expect at least three diploid species had participated in
their origin. The number might be even higher because the point of awn insertion to
the middle of the lemma in *A.magna* and *A. insularis* is epistatic to the awn insertion
at the lower quarter of the lemma in *A. murphyi*, which would requires two diploids
with the same phenotype just for the creation of that tetraploid.

2.6 The Missing Diploid Progenitor of Section Denticulatae

Even if, as hypothesized above, the three tetraploids evolved through genome
reshuffling of an ancestral tetraploid species, or in one of the three, this species has
apparently two diploid progenitors. As already pointed out, none of the known
diploid oat species can be regarded as the progenitor of the tetraploid and hexaploid
species with denticulate lemma tips. Although these putative progenitors are yet to
be discovered, the general features of the spikelet and soil preference of at least one
of them must be similar to those of the three tetraploids and to the *sterilis* type of the
hexaploid oats. Most probably it has a relatively large and coarse V-shaped spikelets which sheds as one unit at maturity, with awn inserted at the lower one-third to
one half of the length of the lemma. The soil in its habitat should be crumbling clay.
No other diploid oats grow in such soil type. This kind of habitat has been arable
land for centuries if not millennia and the hypothetical diploid could have vanished
long ago. What then is the chance of finding it at present?

During the last 50 years seven new oat species have been discovered, three
diploids and four tetraploids. Two of them, *A. magna* and *A. damascena* were
discovered accidently and another two, *A. canariensis* and *A. agadiriana*, as a
result of examining herbarium material. The rest of them, *A. murphyi, A. prostrata*
and *A. insularis* were found and identified according to ecological considerations.

What can past experiences in the discovery of new oat species tell us about the
chances of eventually finding the missing diploid oats? It is difficult to believe that
they will be discovered accidently. Although two new oat species were so discovered, it would not be wise to count on luck for future discoveries. Searching
herbarium material is also not promising in this respect. The two species, in
addition to ssp. *atlantica*, which were brought to light by herbarium surveys of oat
material were collected long ago and it is unlikely that more new species are
hiding in such material.

Three of the seven new oat species were found by paying close attention to the
ecological components of their habitats. In their floristic explorations botanists
stop from time to time along their excursion route to collect herbarium specimens
and to record the plant species in that particular site. Those who are concerned

with plant genetic resources also stop along their way to collect seeds from specific wild or cultivated plants. There are many, and sometimes strange, reasons why plant collectors stop in one place and not in another. As already pointed out, the discovery of A. *magna* happened because Dr. Hayes had stopped near a roses plot to take a photograph and in passing he collected some spikelets in the road ditch which he believed to be A. *sterilis*. Experienced collectors usually select sites for collecting plants or seeds where they notice an ecological change in the habitat. The change could be gradual, as elevation, or abrupt as a change in soil type, bed rock, or vegetation. Different habitats, even if they are in the same general area, are potential sites for novel genetic diversity and even new species. The first site where I collected A. *prostrata*, without knowing that it was a new species, was in a rather abrupt change from alluvial valley soil to metamorphic hills which created a dry habitat. Both A. *murphyi* and A. *insularis* were collected, again, before I knew that they are new species, because I was looking for habitats with uncultivated crumbling clay soil where I assumed that it is the soil type on which they may grow. In searching for additional new oat species one more point must be emphasized. This is an absolute acquaintance of the explorer with the range of morphological variation in the genus and the diagnostic characteristics of all the species. When I first came across A. *murphyi* and A. *insularis* in the field I noticed the peculiarity of their spikelets which suggested to me that they required further study, which indeed turned out to be rewarding.

2.6.1 The Suspected Geographic Area Where the Missing Diploids May Occur

The homeland of A. *magna, A. murphyi*, and A. *insularis* is North Africa with isolated outposts in Southern Spain and Sicily. This could give us a strong indication for the geographical area where the hypothetical diploids may be found. Serious effort to find them is dependent on the availability of detailed soil maps of the area to be explored. Such maps are available for Tunisia but I was unable to get similar maps of Morocco and Algeria. In both Tunisia and Morocco a major problem of exploring the missing diploids is overgrazing and the search would necessarily be confined to protected areas with appropriate soil types. Unfortunately, such areas are quite rare in these countries. In Algeria the situation might be the same but I have no first-handed information of this.

There is no information whether any of these tetraploid species grow in Algeria. I am not aware of any systematic search and collection of wild oats in this territory, but it seems a promising target for exploration of wild oat species. Primarily, because Algeria is located between the populations of A. *magna–A. murphyi* in Morocco and A. *insularis* in Tunisia. If these three species would be found in Algeria, perhaps the missing diploids might be there as well.

2.7 Oat Domestication

Two oat species gave rise to the domesticated forms, the common oat and the slim, or sand oat. In several publications the Ethiopian oat, *A. abyssinica*, is also referred to as a domesticated form but as mentioned before (Sect. 2.3.2), although it contains the domesticated syndrome it is never grown purposely as a crop plant. It occurs only as a tolerated weed in barley fields in Ethiopia. In rainy years the farmers weed it out but in dry years it is harvested, threshed, consumed, and sowed together with barley. The establishment of this oat is a living example of the process of small grain cereal domestication. As *A. abyssinica* originated from *A. barbata* by adaptation to the practice of barley growing by the Ethiopian farmers, so did the weedy forms of *A. sativa* and *A. strigosa* in Europe.

The place of origin of many crop plants and their distribution from a nuclear area to other territories is detected by presence of plant remains, mainly carbonized seeds, in archaeological diggings, and their dating with the aid of C14 carbon dating techniques. Such methods have revealed, for example, that wheat and barley domestication started before 9000 to 10000 years in the Middle East. In several archeological sites from that period both the wild progenitors and the domesticated derivatives appeared in the same stratum. This is the major indication that man had selected either on purpose or unconsciously the domesticated forms.

Interestingly, spikelets of ssp. *sterilis* are very rare in these old remains. This is surprising because this is the most widespread of the wild cereals in places where wild wheat and barley grow. Furtheremore, the first archeological remains of domesticated *A. sativa*, seeds with fracture at their base, rather than a disarticulation scar, are from Europe and as late as the second and first millennium B.C. Why, then, should humans have favored wild wheat and wild barley over wild oat as food resource? To gain a better insight into this question, I embarked on a field study together with some of my students, in an attempt to collect these three wild cereals in the Upper Jordan Valley where they grew side by side (Ladizinsky 1975a). The group was divided into three teams of three persons each, and each team collected a different wild cereal for 1 h. Wild wheat and wild barley were collected by stripping the spikes and oat by tearing off the spikelets from the panicle. We then noticed that in both wild wheat and barley, seeds of a given spike mature more or less at the same time, while in oat when the upper spikelet is mature and about to shed, the lower spikelets are still immature. Therefore, we selected two sites for collecting, one where dispersal was in progress and the second one prior to dispersal. The collected seeds were then dried, threshed, cleaned, and weighted.

Seed yield per collector per hour was more than double in the sites where the three cereals were at the stage just before seed dispersal, compared with the sites where seed maturity was advanced. The yield was 522, 316, and 253 g for wheat, barley and oat, respectively. The low yield of the wild oat was due mainly to the relatively high proportion of immature seeds in the collected material. Wild oat was inferior also from the storage point of view of the harvested material. While the wheat and barley spikes have arranged themselves in storage according to the spike axis, the oat spikelets occupied twice the storage volume because of their twisting awns. It thus became clear to us that prehistoric humans foraging on wild cereals realized that collecting wild wheat and barley was more rewarding than collecting wild oat. When man started engaging with agriculture these two wild cereals were therefore the natural choice for this purpose.

Oat domestication is completely different from wheat and barley domestication. It seems that humans consciously avoided oats as a food resource in the wild, but he could not prevent them penetrating his cereal field as a weed. Once present in man's fields they were subjected to the sowing-harvest cycles of the cereal crops that they had infiltrated, eventually giving rise to establishment of the domesticated oat. The pattern of oat domestication in Europe is probably the same as that of the Ethiopian oat ssp. *abyssinica* with a major distinction. The Ethiopian oat is not a crop but a tolerated weed in barley fields, whereas the common oat and the slim oat are crops in their own right.

The domestication of the slim oat *A. strigosa* apparently took place in the Iberian Peninsula where large populations of the wild form *hirtula* occur even today. It occupies primary as well as man-made habitats and is a common weed in cereal fields. The seed non-shattering character of the domesticated form of *A. strigosa* is governed by two recessive genes (Jones 1940). Because of this, the domestication of *A. strigosa* probably took shorter time than that of ssp. *abyssinica* where the seed non-shattering is governed by four recessive genes (Jones 1940).

The common oat was domesticated in Central and Western Europe where both *sterilis* and *fatua* types are common weeds in cereals fields and other man-made habitats. Therefore, it is impossible to tell if either of them, or both, gave rise to the common oat. Undoubtedly, gene flow from the wild forms to the cultivated oat has taken place long after the common oat was formed whenever these wild forms have infiltrated the common oat fields.

2.8 Domestication Via Hybridization, the Case of *A. magna*

Avena magna is remarkably protein rich. The range of protein content in the groat is between 23 and 27 % among various accessions, while in the common oat is 14 and 18 %. Transferral of this protein content from *A. magna* to common oat cultivars is not feasible for two reasons: first, protein content is a quantitative trait which is controlled by many genes, and second, the pentaploid hybrid between the two oats is sterile to such an extent that does not enable transferring of such a quantitative trait.

Another option for exploiting the *A. magna* protein content is to try to domesticate it by transferring it into the domestication syndrome of the common oat. The domestication syndrome is a set of several morphological characters which differentiate the crop from its wild progenitors. In seed crops the most important trait of the domestication syndrome is seed nonshattering at maturity, non-brittle spike in wheat and barley, and spikelet nonshedding in oats. Under domestication more mutations have been selected by man, causing the crop to deviate further from its wild progenitor. In addition, man selected traits of quantitative nature which affect yield and adaptation.

In the oat crop, in addition to the spikelet nonshedding at maturity, the domestication syndrome is made of erect growth habit, glabrous lemmas, yellow lemmas, reduction, or even absence of awns. Each of them is governed by a single gene (Ladizinsky 1995).

The idea of domesticating *A. magna* crossed my mind when I studied the cytogenetic relations between this tetraploid species and several common oat cultivars. All of the hybrids were produced when the common oat was used as the female parent, no hybrid seeds were produced in the other cross direction. All the hybrid seeds were smaller than the seeds which resulted from self pollinations and could easily be separated from them.

The hybrids developed normally and exhibited hybrid vigor in the number of tillers, plant height and panicle size. In their spikelet morphology the hybrids were intermediate of their parents. Mature spikelets remained attached to the panicle, indicating the dominance or epistatic nature of the common oat over the seed shedding habit of *A. magna*. Lemma color was brown gray and only partly pubescent, and only one awn was produced in each spikelet, always on the lemma of the lower floret. At this stage I realized that *A. magna* cannot be considered the tetraploid progenitor of the hexaploid oats because of the insufficient pairing between their chromosomes. I was further disappointed by the low pollen fertility of the hybrids between the two, less than 1 %, the anthers were shriveled and did not dehisce. Consequently, the hybrids were self-sterile. However, it was a reasonable assumption that normal female gametes in the pentaploid hybrids should occur at the same proportion as the normal pollen grains. To test this I made a large number of hand pollinations of the pentaploid hybrids with pollen of the parental species in the greenhouse, but no seeds were obtained. Another group of pentaploid hybrids were grown in the field amid their tetraploid parent. In the field the *A. magna* plants were severely infected by the viral disease barley yellow dwarf virus (BYDV) and many of them did not reach flowering. Only slight symptoms of BYDV were noticed on the pentaploid hybrids. Despite the heavy BYDV infection of the *A. magna* plants, some tillers produced enough pollen which enabled cross pollination of the pentaploid hybrids. Altogether 33 seeds were collected from the hybrids. They were identified before maturity by being plumper and had green glumes compared with the empty whitish spikelets glumes of the sterile ones.

Chromosome number in the BC plants ranged from $2n = 28$ to $2n = 35$ but only three of them set some seeds. One plant, designated Aa2 ($2n = 32$), had the highest proportion of stainable pollen grains and produced 47 seeds. Two further BC plants, also aneuploids, produced two and four seeds. About half of the BC plants shed their spikelets, indicating that a single gene controls this character. Chromosome counts in 16 progeny of the Aa2 plant revealed that three had $2n = 28$, the rest were aneuploids. Of the 33 F2 plants, 25 retained their seeds at maturity while eight dropped them as the wild parent, indicating that Aa2 plant was heterozygous to seed nonshattering. Four F2 plants were morphologically very similar to the common oat and were characterized by non-shedding seeds, yellow and glabrous lemmas, and only one awn per spikelet or no awns at all.

To gain a better insight into the process of gene introgression from the common oat to *A. magna*, three domesticated derivatives were crossed with three wild accessions. All the hybrids developed normally. Chromosome pairing at meiosis resulted in 14 bivalents and the hybrids were as fertile as their parents. This indicates that the loci controlling the domestication syndrome in the common oat were transferred to *A. magna* by the regular process of crossing over and the chromosome segments that were involved in this were homologous, or homoeologous in *A. magna* and *A. sativa*. Segregation in the F2 populations of the hybrids between the wild and the domesticated *A. magna* showed that each of the four characteristics of the domestication syndrome, i.e., seed nonshattering, bright lemma, glabrous lemma, and reduced awn size and number are each controlled by a single gene but with different dominant-recessive relations. Thus, seed nonshattering was dominant over spikelet disarticulation, bright lemmas were recessive to dark lemmas and so glabrous lemma to pubescent or hairy lemmas, and the reduction of awn size and number was dominant over awn formation on the lemmas of the lower florets of the spikelet. In one F2 family the four genes were linked to one another but in another family, lemma pubescence was excluded from that linkage group, and in the third F2 family spikelet shedding and awn formation comprised one linkage group and lemma color and lemma pubescence of another group (Ladizinsky 1995).

While selection of derivatives containing the domestication syndrome was encouraging, it was obvious that a lot of breeding effort would be required to bring them to the status of cultivar or at least a useful genetic stock. Several characteristic had to be improved such as weak straw, tough rachillas that made threshing difficult, some sterility (the reason for which I could not pin point), large glumes and susceptibility to powdery mildew and BYDV. *Avena magna* is notoriously susceptible to these diseases but the common oat cultivars which were used in this study were resistant, or at least tolerant. The F1 hybrids were also tolerant but I could not find any F2 plant with such tolerance. The solution to that drawback was another hybridization cycle. This time the domesticated tetraploid type was used as pollen donor with the advantage that no progeny would be discarded because of wild parental characteristics. All together four additional hybridization cycles have been attempted. In each cycle an improved tetraploid type was employed for the next cross. The duration of each cycle was 3–4 years and the whole project has taken over 20 years. The end result is a domesticated tetraploid with erect growth habit, better but not enough straw stiffness, smaller glumes, reduced sterility but no improvement with resistance to BYDV and powdery mildew.

Throughout the domestication project the focus was on plant shape but not the protein content of individual derivatives. The idea was that transferral of the protein content would be the last stage by crossing the wild and the domesticated forms, raising large F2 populations and selection of protein-rich domesticated tetraploid oat. Preliminary results show that in such F2 populations domesticate forms with protein content as high as 26 % do exist and growing much larger F2 population is now under way.

2.8.1 *Avena magna Domestication as a Model for Other New Crops*

As far as I am aware, besides *A. magna*, no other wild plant has ever been domesticated by transferring into it the domestication syndrome from another crop plant. To what extent then, can this experimental procedure be utilized for domesticating other wild species with economic potential? To accomplish this, at least two preconditions must exist: (1) the wild species must be cross compatible with a crop plant that would provide the domestication syndrome. (2) The hybrids between them must develop normally and produce some viable gametes. The rest is back crossing and selection. Obviously, back crossing would be much easier with wind pollinated species than those which are self-pollinated. However, it should be stressed again that even if the two preconditions for domesticating a certain wild species are met, the decision of entering the domestication adventure depends upon the nature of the characteristic of the wild species which is the target for the domestication. Traits that are governed by a single gene or a small number of genes can be transferred by introgression to a domesticated relative. Only when the character of the wild species is genetically complex or governed by many genes, the domestication adventure may be justified.

2.8.2 *Production of Synthetic Hexaploid Oat by the Domesticated A. magna*

Besides its scientific interest, production of synthetic hexaploid oat combining the newly domesticated *A. magna* and some diploid species may have some economic value. For example, production of such a synthetic hexaploid oat together with *A. strigosa* var. Saia has the potential of being protein rich and BYDV and downy mildew resistant. Such a synthetic hexaploid oat might also be a useful bridge for transferring useful characteristics from other diploid oats to the common oat.

For producing the synthetic hexaploid, var. Saia of *A. strigosa* was crossed, as female parent, with the domesticated tetraploid. The hybrid plants developed normally and when they had 3–5 tillers they were treated with colchicines solution. As a result, more than half of the tillers died or stopped developing and new tillers emerged. At maturity most of the tillers had born empty spikelets but in some, a few seeds were produced. All the plants that developed from these seeds were hexaploids ($2n = 42$).

The synthetic hexaploids developed normally and exceeded both parents in plant height. While the panicle of var. Saia is cylindrical and condensed those of the domesticated tetraploid were shorter and more open. The panicles of the synthetic hexaploid had intermediate appearance between the parents. Spikelets of var. Saia usually bear a single fertile floret with a rudimentary awn whereas the domesticated tetraploid spikelets bore two fertile florets with no awn. In this respect the synthetic hexaploid was similar to both parents; it had two florets per spikelet but only the lower one was fertile. On the other hand, spikelet size was larger than those of its parents. The same was true for seed size but the seeds exhibited various degrees of shriveling with 16 % protein as in var. Saia. After three generations plants with an improved seed/spikelet ratio, 1.7, were selected with much reduced shriveling, but with no improvement of the protein content. In the field, the synthetic hexaploid exhibited crown rust resistance but was as susceptible to BYDV as the domesticated tetraploid, although var. Saia is resistant to BYDV. This indicated that the BYDV sensitivity of the domesticated *A. magna* is epistatic to the resistance of var. Saia to this disease. Such information is important in any attempt to transfer BYDV tolerance from *A. sativa* and selection for BYDV tolerant domesticated tetraploid. Such selection should start only at the F2 of the cross (*A. sativa* x domesticated *A. magna*) x domesticated *A. magna*), and continuing in more advanced segregating generations.

In the triploid hybrids (*A. strigosa* x domesticated *A. magna*), chromosome pairing at meiosis was similar to the *A. strigosa* x *A. magna* cross combination. Most of the chromosomes were left unpaired and the few bivalents were rod shaped with terminal, an end-to-end type of chiasma. Chromosome pairing improved dramatically in the synthetic hexaploids where most of the chromosomes were associated as bivalents with occasional one and rarely two quadrivalents resulting in a high number of chiasmata per cell. Restoration of chromosome pairing in this case can be attributed to the potential for preferential pairing of *A. strigosa* genome in the presence of alien genomes (see Sect. 2.3.7). In the third generation, 21 bivalents were regularly formed at meiosis and univalents or quadrivalents were not observed.

Hybrids between the synthetic hexaploid and common oat cultivars were easily obtained and in both cross directions. Meiosis in these hybrids was irregular with up to 11 univalents and 4 quadrivalents per cell and were therefore self-sterile.

Chapter 3
Collection of *Avena* Wild Genetic Resources

Abstract Loss of genetic diversity, a side effect of the green revolution, has provoked national and international efforts to collect and preserve the genetic diversity of crop plants and their wild relatives. In 1984, the *Avena* Working Group was established to coordinate the collection, preservation, and documentation of oat genetic material. Collecting wild relatives of oats, as of other crop plants is entirely different from collecting cultivated germplasm. Successful collection of wild relatives requires knowledge of botany, ecology, soil science, and geology. Careful preparation of the collection trip is essential for collecting material that potentially represents the genetic diversity in the wild. The field work comprises several steps which include identification of sites where the target species is expected to grow, actual collection of seeds, and recording the information of both the site and the accession.

Keywords Genetic diversity · *Avena* working group · Preparing collection trips · Collection in the field · Documentation · Evaluation · Utilization

Development and distribution of high yielding and widely adapted cereal varieties together with introduction of irrigation technologies and artificial fertilizers during the 1960s and the 1970s of the previous century have made dramatic changes to food production in India and other developing countries. This move is now thought of as the Green Revolution. The immediate effect was a sharp increase of cereals yields with remarkable benefits for the farmers and removal of the fear of famine that was common in the Indian subcontinent. The drawback was massive abandonment of the local varieties and land races which had been used by farmers for generations because they were inferiors to the newly introduced cultivars. Land races are usually locally adapted but are not high yielding.

In response to the growing alarm over the rapid loss of crop diversity, the International Board of Plant Genetic Resources (IBPGR) was established in 1974. Its mandate was to coordinate an international program for plant genetic resources that included organizing collection missions as well as building and expanding

Gene Banks at national, regional, and international levels. For many crops, working groups were established to recommend actions which were necessary for collecting and conserving genetic resources of the crop and its wild relatives.

3.1 *Avena* Working Group

In 1984, I was invited to participate in the first meeting of the *Avena* Working Group (AWG) which was held in Izmir, Turkey. The recommendation of that meeting was to initiate a survey of common oat germplasm in different countries, mainly in Europe and to collect seeds of wild relatives in the Canary Islands, Spain, and Morocco with emphasis on the species *A. canariensis, A. prostrata, A. magna* and *A. murphyi*. I was asked to organize and conduct this collecting mission. My only reservation was that as an Israeli I could not enter Morocco. They promised to provide me with a UN passport for this purpose. Another suggestion that I made was to take that opportunity to train a young oat researcher to become familiar with all the aspects of collecting wild relatives. Mike Leggett from UK was suggested and approved. I had previous experience in collecting *A. canariensis*, *A. prostrata*, and *A. murphi* but not with *A.magna* because I have never had the opportunity to collect in Morocco. As soon as I could, I started collecting information about Morocco, geography, ecology, pedology, geology, and vegetation. In addition, I listed all the sites where *A. magna* was collected or observed.

About a month before the beginning of the mission I was informed that IBPGR was unable to provide me with a UN passport and I would not be able to go to Morocco. As Mike Leggett was not yet ready to go alone to Morocco I suggested to the IBPGR secretariat that they appoint an additional young oat worker who would join the mission. Furthermore, both of them would spend a week in Israel to get preliminary experience in collecting wild oat species. My suggestions were approved and Per Hagberg of Sweden was selected.

The week with Mike and Per in Israel was highly intensive. In a very short time they had to see, learn, and absorb the various aspects of collecting wild oat species in their natural habitat; to recognize how various ecological parameters combine to create the habitat of the various oat species and to detect these species among other natural grasses, including more common oat species; to identify the main morphological characteristics of the various species; and how to separate the diploids ssp. *wiestii* and. *hirtula* from *A. barbata*. In addition, they had to learn how to determine soil type and to distinguish between hard limestone and chalk and the landscape they create. I realized that one week was not really sufficient to come to grips with all this information, but I realized that the training would be continued in the real collection trip.

A week after they left Israel we met again in the Canary Islands to collect *A. canariensis*. We spent most of our time on two islands, Fuerteventura and Lanzarote. As with the other Canary Islands, Fuerteventura and Lanzarote resulted from volcanic eruption. In Fuerteventura, the landscape is mostly flat with barren hills, annual vegetation, and few shrubs. The island is poorly inhabited and agriculture is extremely limited. The economy is based on tourism. The weather is pleasant but rainfall is low, about 120 mm per year mostly in the autumn and in the beginning of winter. *Avena canariensis* occurs there on the hills and in flat areas in undisturbed niches, sometimes together with ssp. *sterilis* and *fatua* .

Lanzarote is hillier than Fuereteventura and agriculture is more widespread, particularly at the higher elevations where they grow vegetable, legumes, corn, and even grapes in vineyards, all of which are rain fed, again about 120 mm annually. In Lanzarote, *A. canariensis* seems to be more widespread and the individual populations are larger compared to Fuerteventura.

From the Canary Islands we continued to southern Spain to the area where *A. murphyi* is native. We had two surprises there, the season was early and the occurrence of this species was much more limited compared to what I had observed in 1969. Some of the grazing plots had been turned into arable land and the tetraploid oat had vanished. Nevertheless, we were able to locate a number of populations for seed collection but the seeds were immature. We agreed that Mike and Per should continue to Morocco and I would stay another week in Spain to collect the seeds. Before they left I handed over to them all the information that I gathered about Morocco and the locations where *A. magna* was collected or observed. I also indicated to them my educated guess, or gut feeling, with regard to areas in Morocco where new species might occur. I mentioned the area south of Tangier where alluvial soil is common and could be an appropriate habitat for *A. murphyi* and the area between Tiznit and Tafraut in southern Morocco, where the climate is dry and sandy soil is rather common. The achievements of Mike and Per in Morocco were astonishing. Any experienced oat collector would have been proud of collecting what these two young collectors had accomplished. They found *A. murphyi* south of Tangier, *A. damascena* and *A. prostrata* which previously were unknown in Morocco. In Tiznit-Tafraut-Agadir area they found two new types which they believed were new species (Leggett et al. 1992). However, several months after they returned from Morocco these wild oats were published as *A. atlantica* and *A. agadiriana* by Canadian oat workers who collected them 2 years earlier (Baum and Fedak 1985a, 1985b). Their mission to Morocco resulted from information that had emerged following examination of herbarium material that directed them to that specific part of Morocco.

I spent another week in southern Spain but the *A. murphyi* seeds were still immature. I had to find someone that would collect the seeds at the appropriate time because I could not afford to stay in Spain any longer. In Cordova I knew Dr. J.I. Cubero from his publications on broad bean. At that time I was studying the cytogenetic relations between broad bean and several wild *Vicia* species in order to test the hypothesis that the broad bean was domesticated from one of them, which I refuted (Ladizinsky 1975c). I drove to Cordova to seek help for collecting the *A. murphyi* seeds from Dr. Cubero. Unfortunately, he was unable to do it but he knew an agronomist working in the Andalusian extension service in Jerez de la Frontera who might be able to help. He called him and later we met. I showed him the populations of *A. murphyi* and how to distinguish them from ssp. *sterilis* and he promised to do the collection. After 3 weeks I got the material, all the samples were of *A. murphyi*, except one, which was ssp. *sterilis*.

In the early 1990s the Moroccan government allowed Israelis who were born in Morocco to visit the country. Shortly after that any other Israeli citizen was welcome providing that they entered and left Morocco in an organized group. This was my chance to enter Morocco and examine wild oats in that country. I arrived with a group in Tangier and left it immediately after landing and met them 10 days later in Casablanca airport to leave Morocco. During my visit I rented a car and drove to see the habitats of *A. magna, A. murphyi, A. atlantica* and *A. agadirana* sites. My first stop was at the outskirts of Tangier, a site of *A. murphyi*, which was discovered by Mike and Per. The soil was similar to that in Southern Spain and many plants had black spikelets, a character which I have not seen in the Spanish populations. About 5 km from Tangier on the road to Casablanca, at the junction to Cap Spartel, I noticed a large population in an area of more than one hectare which apparently once had been cultivated, where *A. murphyi* formed nearly a pure stand. Another isolated population of *A. murphyi,* was seen near Beni Slimane, east of Casablanca, on the same soil type as near Tangier. The site was amid hills of sandy loam soil where the cork oak was particularly common. In a subsequent visit to the habitat of *A. murphyi* in the Tangier area,some of the populations no longer existed because of urbanization and the future of the remaining of the populations seemed rather gloomy.

My main interest however, was the habitat of *A. magna*. I had realized that this species is not as common as I had thought and was restricted to the edges of cultivation and roadsides where it was protected from grazing. I could rarely find it in open habitats, for example in the south and east of Ouazzane, again because of overgrazing which is a severe problem in Morocco. Rarely *A. magna* was found in cultivated fields, as along the road from Fez to Ouazzane, where individual plants occasionally reach the height of nearly 2 m. The soil in the *A. magna* habitat is crumbling clay, in color ranging from reddish to brown black. This soil type is almost totally under cultivation and that is why *A. magna* is so rare.

3.2 In situ Conservation of Wild Oat Species in Southern Europe

In 2008, I was invited by Andreas Katsiotis to participate in a meeting in Athens on in situ conservation of wild oats. During the meeting it became apparent to me that there was a plan for in situ conservation of wild relatives of several crops in Europe and oat was one of them. We were supposed to select the oat species for that program and the particular populations for each species.

The common way of conserving genetic resources is by collecting seeds of the target species and putting them into cold storage for a number of years, which is termed ex-situ conservation. The advantage of this method of conservation is that a large number of seed samples from the entire distribution and diverse habitats can be collected, stored, and evaluated when needed. The disadvantage is that after several years in storage each sample has to be regenerate to ensure continuing seed viability. Another drawback is that the stored samples are no longer evolving and even might lose some of their variability if some genotypes require more frequent regeneration time than others.

Another approach for conserving genetic resources is by in situ conservation. In this case, the entire population is conserved by providing appropriate protection and management of the site. The advantages of in situ conservation is that it is cheaper compared to gene bank operations; seeds can be collected when they are needed and the population as a whole continues to evolve in response to new selection pressures. The main disadvantage is that it is impossible to apply this approach on the entire distribution of the target species and usually it can be attempted on one or only a small number of populations. This limitation must be borne in mind also when attempting in situ conservation for oat wild genetic resources.

In the Athens meeting, I expressed my opinion that selection of wild oat species for that program should be made according to their potential usefulness as gene donors to the cultivated oats, their rarity, and if they are native to Europe, even if they are members of the tertiary gene pool of the cultivated oats and at the moment cannot be regarded as effective gene donors. The recommended species were *A. ventricosa* in Cyprus, *A. insularis* in Sicily, *A. murphyi* in Sothern Spain and wild *A. strigosa* in Spain and Crete. *A. ventricosa* is rare on global scale with a few and disjunct populations from Baku in Azerbaijan to Oran in Algeria with only one location in Europe, in Cyprus. *A. insularis* and *A. murphy*i are members of the common oat's secondary gene pool and occur in restricted areas with unique soil types in Sicily and Spain respectively. On the other hand the wild forms of *A. strigosa* are much more common than the other three species. However, they are members of the primary gene pool of the domesticated *A. strigosa* and it would be worthwhile conserving populations from the western and eastern fringes of this oat in Europe.

My proposal was accepted but it was necessary to check the suitability of the proposed populations for in situ conservation. It was also essential to examine these populations in the field and to check their suitability for this program, before making the final recommendation. I also indicated to A. Katsiotis that I would be available for the field study if he thought it appropriate, which he did. We agreed that we would visit Cyprus and Crete in spring 2009 and Sicily and Spain in 2010.

During the preparation for the field studies in Cyprus and Crete I tried to get as much information as I could on the distribution and acreage of protected areas on both islands. The reason was that if the target species occur in protected areas, of any kind, its conservation is almost automatically guaranteed. It looked quite promising for Cyprus but not so for Crete, where the number of the protected areas was much smaller and they were outside the range of habitats where wild *A. strigosa* usually grows.

In Cyprus we were accompanied by Angelos Kyratzis, an ECPGR National Coordinator. Within 2 h after landing, we detected nearly all the wild oat species which are known in the island, including *A. ventricosa*. Particularly extensive populations of this species were observed in protected areas on the east side of Lake Larnaca, on the slopes of hills near Agios Sozomanos, and in Athalassa National park near Nicosia, all on calcareous bed rock. Smaller populations on the same soil and bedrock were seen around ARI near Nicosia, near Kutrafas, close to Flasou, and east of the British autonomous military base, most of them within protected areas. *A. ventricosa* was found also on basalt soil, mostly in pine groves where it forms rather small populations, on the road from Dilikipos to Lythrodontas and between Mathiatis and Analiontas and on Mathiatis-Sia road.

In several sites wild *A. strigosa*,(*hirtula* type) grew adjacent to *A. ventricosa* and also in areas where it was found together with *A. sterilis*, mostly in protected areas. In such sites the *eriantha* type of *A. clauda* was also occurred.

The visit to Cyprus indicated to us that this island is most appropriate for in situ conservation of *A. ventricosa* and the *hirtula* type of *A. strigosa* because of the wealth of their populations, and the diverse ecological niches they inhabit and the extensive protected area network existing there. Consequently, we agreed to recommend Cyprus, other than Crete, also for in situ conservation of the *hirtula* type of *A. strigosa*.

Andreas Katsiotis and I arrived in Sicily in early May of 2010 at the right time for seed collecting. We were accompanied by Ferdinando Branca, the chairman of the Brassica Working Group, and we drove straight to Gela area, the homeland of *A. insularis* and arrived at Lake Comunelli. We stopped at the site where 15 years earlier I had discovered for the first time this species. Some new road constructions had been made there, but I could still recognize the site and found several *A. insularis* plants surrounded by much taller *A. sterilis* stand. We then climbed up the uncultivated hills which were covered by herbaceous vegetation, mostly annuals, where pure stands of *A. insularis* were prevalent. Later, we came to the other side of these hills where large-scale afforestation had been attempted 20 years previously with eucalyptus and pine trees. Under the shadow of the trees not a single plant of *A. insularis* was found but they were common in the open spaces between the trees, together with the perennial grass *Lygeum spartum*. The eucalyptus trees apparently are not adapted to the soil type at that place and many of them had died after several years. In such clearances *A. insularis* became the dominant plant.

On our request, F. Branca brought the soil map of Sicily, Carta dei Suoli della Sicilia, by Giovanni Fierotti, published by Palermo University in 1988. The map comes with a booklet of detailed descriptions of the various soil types of the island. The soil type around Lake Comunelli is designated as association no. 13 with the following characteristics:

- Modified—regusuoli
- Soil taxonomy, USDA—typic xerorthentes
- Soil taxonomy, Unesco—eutric regosols
- Compos %—35 to 55
- Phases-eroded—gully

- Inclusions—vertisol
- Substrati—clay
- Depth—shallow to moderately deep
- Texture—fine-medium
- Morphology—hilly
- Altitude—500 to 900 m
- Slope—Slopping to moderately slopping.

According to the map and the booklet, this soil type is the most extensive association in Sicily, 13.38 %, covering 344,200 ha. It makes up a large part of the clay hilly area and it is found most extensively in the provinces Agrigento and Caltanissetta. It can be found from sea level to a maximum elevation of 1,500 m.

The following day we traveled to the area near Mt. Bubonia, southeast of Caltagirone where 12 years back I found another population of *A. insularis*. Unfortunately, I could not find that particular site, perhaps we took the wrong track or missed the turn. However, on the road P96 from San Cono to Gela, about 3 km north of the junction with road P190, I noticed an earth road leading to an area planted with eucalyptus. After about 1–2 km on that path we found a large population of *A. insularis*, particularly on one hill with no eucalyptus trees. Then we realized that this hill had been planted with pine trees without removing the lush annual vegetation. The choice for pine apparently was made because they are more drought resistant than Eucalyptus. Nevertheless, if the pine planting turns out to be successful, it will eliminate the wild oat population there because they cannot survive under the shadow of the pine trees.

After consulting the soil map we decided to check another area, about 50 km to the northeast, about 3 km south of the Catania-Palermo highway on road P 102 between Cinquegrana and Sferro. About 2.5 km. from Cinquegrana at about 500 m south of the road we saw a cliff with soil type as association 13 covered with herbaceous vegetation with a profusion of *Lygeum spartum* plants and when we got closer we found an extensive population of *A. insularis* thriving there. Checking other sites with similar soil type further north yielded no additional populations of *A. insularis*.

The three sites of *A. insularis* in Sicily are not in the Natura 2000 scheme of protected areas. However, two of these sites are designated protected areas by the Forestry Department. Applying in situ conservation for *A. insularis* in these sites must be made in close collaboration with representatives of the Forestry Department which should be fully informed about the morphology and ecology of this oat species. It should be stressed that *A. insularis* grows in open habitats and planting trees there, particularly pine, would eliminate the oat populations instead of protecting them. While I have been involved in the exploration phase of the in situ initiative, I have no idea if and how it would be implemented in Sicily.

Any of my visits to Sicily yielded additional populations of *A. insularis*. I then felt that a thorough survey of this species in its unique habitat may give a reliable indication regarding its distribution in Sicily. From the University of Palermo I obtained a copy of the soil map and in spring 2011 I went to Sicily for a week.

I examined nearly all the areas where soil type no. 13 was indicated on the soil map. As expected, that soil type is almost entirely under cultivation, even on remarkably steep slopes. Altogether I found additional nine populations of *A. insularis*. As the previous populations, they were found on uncultivated land in the southern parts of the Island. All of them were rather small and restricted. So, *A. insularis* can be regarded a rare species in Sicily and it is worthwhile protecting it.

The visit to Spain for monitoring the situation regarding *A. murphyi* had some initial difficulties. It appeared that the wild genetic resources issue has been decentralized in Spain and each province has now its own policy and strategy. The ECPGR application for our visit to monitor *A. murphyi* populations in Andalucía was turned down and only after intensive correspondence we were allowed to come. In Spain we were accompanied by Pedro Garcia. I was amazed to find out how rare *A. murphyi* had become in its natural habitat in the southern tip of Spain. Since 1969 I have visited that area several times and the gradual vanishing of this wild oat was a matter of great concern to me. Along roadsides only few plants could be found and in most of the grazing plots it was totally absent. In some cases it was absent in one plot but common in a neighboring plot. It indicated to me that this must be a result of different management and grazing practices. Then we came across one plot where *A. murphyi* was particularly common, nearly in a pure stand. We wanted to meet the landowner and ask him about the grazing practices he followed. In a small restaurant next to the road they told us his name, Pedro Moyda, and pointed out his house on the top of the hill. After sometime we met him and his son, and walked together to his pasture land, where Pedro Garcia showed them the morphological difference between ssp. *sterilis* and *A. murphyi*. They were astonished to learn that there were two wild oat types on their land. When we asked about the grazing practice they followed, they said that they kept the cattle out of the plot when the oat plants start flowering and brought them back only after seed dispersal. Pedro Moyda invited us to his house and told us about life in that part of Andalucia. He was 85 years old, still active and was born in the same house where he now lives. As long as he could remember, they had kept cattle more or less in the same way they were still doing today. In his youth they used to ride on a donkey to Tarifa, half a day in each direction, to buy provisions. Electricity had reached them only 5 years ago and water was still conveyed from a nearby spring on the back of mules. The unexpected meeting with P. Moyda and the educational conversation was enlightening.

Even before coming to Spain I had realized that in situ conservation of *A. murphyi* might be difficult because currently it grows in grazing plots and announcing these grazing lands as protected areas for the sake of conserving that tetraploid oat species was unthinkable. On the other hand, the landowners were not aware of its existence on their property and its contribution to the productivity of their pasture. It then became obvious to me that by proper grazing management both the cattle growers and this wild tetraploid oat would benefit. The scheme seemed rather simple, keeping the cattle out of the plots from oat flowering to seed dispersal every second or third year, thereby ensuring production of sufficient amount of oat seeds and storing them in the ground to sustain extensive oat populations.

Before leaving Spain we met in Seville a number of officials who were dealing with genetic resources and we shared with them our observation and thoughts regarding ways of sustaining *A. murphyi* populations in their natural habitat, by educating the farmers about the presence of this oat on their land and the benefit they might gain by applying appropriate grazing management.

One of the officials in Seville showed us a map indicating sites where they monitor *A. murphyi* populations. One of these sites was about 60 km north of the main distribution of this species and we were much interested in visiting that site which is a part of a protected natural park, between Alcala Los Gazules and Ubriqe, close to Aula de la Naturaleza El Picacho. The actual site was at the bottom of a gorge about 1.5 km from the main road and leading to it was a path blocked by a gate. The path cut through the park, where remnants of olive plantations and scattered cultivation could be seen, with a few cows roaming the area. At the bottom of the gorge there were a number of clumps of *Gladiolus* sp. in blooming and loose stands of ssp. *sterilis*. Then we noticed a number of *A. murphyi* plants, altogether not more than 15 in all. I wondered how the park rangers had detected such an inconspicuous small population in that remote gorge, and assumed that they probably monitored the *Gladiolus* population and recorded the accompany plants on the site. Much credit is due to them for being able to distinguish between ssp. *sterilis* and *A. murphyi* and to add to the inventory of this species additional population, about 60 km from its main distribution range.

On our return to the main road we noticed a few plants of *A. murphyi,* just opposite the gate on the other side of the road. This was really quite strange because the habitat was so different from what we had seen at the bottom of the ravine and in the Tarifa area. We knocked on the door of a nearby house in order to talk to the landowner of that farm regarding the history of the farm and to try to get some hints about the origin of the tiny population of *A. murphyi* there. Unfortunately, there was no one at home and we were left with two theories (1) the seeds of *A. murphyi* were somehow brought to that site by humans or, (2) we still have a lot to learn about the distribution and ecology of this species in Spain. I prefer the first option but further study may prove otherwise.

Besides seeking *A. murphyi,* the mission to Spain was aimed at finding areas for in situ conservation of the *hirtula* type of *A. strigosa*. In previous visits to Spain I came across this wild oat in a number of sites including the area near Seville and along the road to Portugal via Huelva. It grows there mainly on sandy and sandy loam soil. In that general area there is a famous natural reserve, The National Park of Donana, a wetland which is closed to the public, but on the other side of the road there is another protected area named Acebuches which is open to visitors. In that area we found sustainable populations of the *hirtula* type together with *A. longiglumis*. So, unlike the in situ conservation of *A. murphyi*, which seems problematic at the moment, in situ conservation of the *hirtula* type of *A. strigosa* seems to be well in hand.

3.3 The Methodology of Collecting *Avena* Wild Geneitc Resources

For any crop, oats included, collecting wild genetic resources is profoundly different from collecting domesticated germplasm. The reason is that for collecting cultivated oats germplasm the collector has to be able to identify this crop in the farmer field, whereas for wild genetic resources it is important to be acquainted with the morphology and the diagnostic characters of each of the wild relatives to be collected. The collector must also know a great deal about their habitat and the factors affecting that habitat such as soil, bed rock, and have the botanical knowledge for identifying the plant communities which the target species is likely to grow.

Collecting wild relatives as genetic resources comprises two phases, a preparatory and a field work phase.

3.3.1 The Preparatory Phase

Adequate preparation for collecting in the field is the key to a successful mission. It includes gathering information regarding the target species, its geographic distribution and ecological preferences. That information can be obtained from regional floras and herbarium material, the latter is particularly important and because of several aspects:

1. Examining the plants of the target species and identifying their diagnostic characteristics.
2. Detecting the site where it was collected. As already mentioned herbarium sheets usually come with written information about specific specimen including the site where it was collected. Useful information includes distance and direction from the nearest settlement, or other conspicuous land marks, which should be taken into consideration in planning the excursion route.
3. Collecting date. There are several reasons why this information is helpful. If the collection is recent this suggests that the target species still grows on the same site. When coupled with the degree of seed maturity of the specimen it indicates the proper time for seed collection at that particular site. If specimens collected from the same site in different years, the degree of seed maturity in them may provide some ideas about time variation in seed maturity.
4. Ecological remarks. Written information about the herbarium sheet usually includes notes regarding soil and bedrock of the site and the attendant plant community there. This ecological information is particularly important in attempts to find the target species in places from which no herbarium material or other information is available.

3.3.2 Planning Route and Time of the Field Trip

Before deciding on the route and the time of the field trip additional knowledge must be obtained on the target area. It should include information about soil, bedrock, and vegetation and obtaining maps that detail these parameters is particularly useful. The actual route is determined according to the information gathered during the preparatory phase. The route should pass in areas where the target species is reportedly present or where it is likely to grow. A general recommendation is to avoid as much as possible main roads and high ways because only weedy species may grow there. Secondary and earth roads are preferred because they usually cut through undisturbed primary habitats where most wild oat species grow.

Appropriate determination of the time of the field trip can make the difference between success and failure. Seed dispersal of wild oats occurs right after maturity and may last 2 weeks or so. By arriving too early the seeds are likely to be immature, and arriving too late the seeds are likely to have already shed and might be buried in the soil. Variation in maturity time is directly affected by the amount and distribution of rain during the winter and collecting information regarding these parameters is essential.

Seed collection in foreign countries, or even just botanical survey, usually involves acquiring the appropriate permit. It is important to verify whether such a permit is necessary in any particular country and how it can be obtained. Collection trips funded by international organizations are usually approved only after presenting documents of collaboration with local institutions or scientists. Such collaboration is sometimes useful, when the collaborator is familiar with the vegetation of the country, for example, but in some situations it can be a burden.

3.3.3 The Field Work

The field work comprises several steps, selection of sites for seed collection, identification of the target species on the site, assessment of the population size, preliminary identification of the major ecological components at the site, seed collection, and filling in the collection forms.

3.3.3.1 Site Selection

In selecting sites for collecting oat wild genetic resources two situations can be envisaged: the site has already been selected during the preparatory phase or, that no previous information is available. The second scenario is more problematic because the collector has to decide where to stop and start searching for the target species. Although oat plants are rather conspicuous and can be spotted even by driving 60 km/h, but along the roadside there are mainly weedy species which are of secondary interest.

In my experience, selecting sites according to ecological parameters is much more effective than blind selection, because most oat species are adapted to specific habitats and environments. Change in ecological parameters along the route, such as soil, bedrock, topography, and plant community are good reasons to stop and start searching for the target species. When the excursion route runs for considerable distances through more or less the same environment with similar ecological components, it is useful to collect every several kilometers.

3.3.3.2 Brief Identification of Ecological Factors on the Site

For an experienced collector, identifying these parameters may only take a few minutes. It includes soil type and bedrock formation, topography and plant community. It is also desirable to evaluate ecological variation within the site as sub niches and it would be appropriate to include them in the collecting area on the site.

3.3.3.3 Identification the Target Species on the Site

Even when the route taken is on secondary and earth roads, target species only rarely are found next to the road. Usually some walking and hiking is necessary before finding the target. Sometimes, the habitat seems adequate but not a single plant of the target species is to be found, which is rather frustrating. For the sake of efficiency and time saving, it is useful to leave it after 10–15 min. if you do not find what you are looking for and search, for another site.

Always, the target species is a member of a plant community and on the same sites many other species, annuals, and perennials will grow as well, as other oat species, and they may be more predominant than the target species. The target oat plants may grow in a pure stand, in clumps or as solitary plants. When the latter is the case, finding the first individual is sometimes the "turning point", because after that you may see it everywhere.

It is worth reiterating that appropriate recognition of the diagnostic characteristics of the target species and the range of morphological variation in the whole genus are indispensible for the identification of species in their natural habitats, and off types that may occur there.

3.3.3.4 Population and Site

Populations of plants, including wild oats, can be limited and local or extensive with no definite borders. The site where the seed collecting is exercised may be identical in size to that particular population, but more usually, is just a small part of it. The area in which sampling of the wild oats is taking place and its dimension might depend upon ecological parameters. The more ecologically diverse the site is the larger should be the sampling area.

3.3.3.5 Seed Collecting

The actual seed collection involves several theoretical and practical issues such as how many seeds to collect, from how many individuals, and what should be the distance between the sampling points and their distribution on the site?

Collecting wild genetic resources for ex-situ conservation in oats and other crop plants is usually undertaken for future needs that at the time of collection is difficult, if not impossible to predict. In fact, collecting wild species for ex-situ conservation is often a blind collection of genetic diversity that practically we know nothing about. What is then the likelihood that a gene or allele controlling a characteristic that would be needed in the future will be included in the seed sample collected at a particular place? While it is impossible to provide a straightforward answer to this and other relevant questions, it is legitimate to consider it and for gaining some insight into the general pattern of allelic distribution in natural populations.

Let us assume three situations:

1. The allele in question is widely spread. It occurs in every population and at high frequency. Such an allele is expected to be found in nearly every seed sample.
2. The allele is rare globally but may be common locally. It occurs only in a small number of populations, but there it would be rather common. The frequency of that allele in a particular population would imply the number of individuals from which seeds have to be collected for assuring that the particular allele has been included. When the frequency of the allele in question is 5% or more, collecting seeds from 50 individuals would give such an assurance.
3. The allele is rare locally and everywhere.

In collecting wild genetic resources, alleles that are rare globally but common locally are of greatest value. That kind of distribution, most likely, reflects adaptation to a specific environment resulting from unique local selection pressure. Because local environments are crucial for the occurrence of the target allele, some experts recommend that field work should increase the number of sites but reduce the number of individuals from which seeds are collected.

Oats are self-fertilizing and individual plants are highly homozygous for most of their genes. Wild oat plants usually disperse their seeds close to the mother plants. The result of this is that at each particular spot most of the plants are sibs and share the same genotype, but other genotypes may occur in other spots. Therefore, it is recommended to collect seeds from individual plants which are distant from one another by about 10 m. This gives collecting track of about 500 m, but often it is longer because wild oats are not homogeneously distributed in the site and the attempt to collect seeds from plants growing in the various ecological niches of the site.

Such ecological niches are soil type, deep or shallow or stony soil, open habitat, or under the tree canopy, different slope aspects at the site, etc.

Since seeds of the same oat plant are genotypically highly similar, only a small number of seeds should be collected from each individual but be kept in separate bags. All the 50 bags are then put in a larger bag which would be labeled with the site number and other relevant information regarding that material. Another option is to collect a single dispersal unit from each plant and even by placing them all in one bag they will keep their identity and reflect the genetic diversity at that site. While this kind of packing reduces dramatically the volume of the collected material, it requires multiplication by the gene bank, before distributing it to other institutions, or evaluating for various purposes. Handling an accession which is composed of potentially 50 different genotypes poses problem in gene banks which usually regard an accession as genetically rather uniform. This may be correct for cultivated material but not for wild accessions collected the manner described above.

3.3.3.6 Documentation

The value of collected material is considerably reduced if it is not accompanied by written documentation, a collecting form about the sample. Collectors may use different types of forms but they include more or less the same items and information. The minimum information required, known also as minimum passport data, comprises:

1. Collector name and Institution, date of collection, and collection number.
2. Botanical information, species and subspecies names, and whether or not confirmation is needed.
3. The site, country, province and exact location, coordinates, altitude, nearest landmark and direction. Pinpointing the exact site by GPS is highly recommended for facilitating return to the site, if necessary.

   Additional desirable information may be:

4. Ecological remarks such as, topography, soil, and bedrock, the major components of the plant association and degree of disturbance by grazing, cutting, cultivation, and urbanization.
5. Demographic details such as, occurrence in the site, common or rare, uniform stand or patchy distribution, degree of variation and of which characters, incidence of pests and diseases manifestation.
6. Details about the collected sample, bulk or from individual plants and how many, and the size of the sampled area.

The value of the additional information may become apparent after evaluating the collected material by relating specific traits of economic importance that would be found, to ecological parameters and genetic diversity to geographic distribution.

3.3.4 *Handling Wild Genetic Resources in Gene Banks*

The main difference between handling domesticated and wild germplasm in gene banks is the special care needed when regenerating the latter. One of the reasons is that seeds of wild species contain germination inhibitors, mainly in the husks and by planting them directly in the field they may fail to germinate and get lost. Therefore, pre-germinating testing of a small number of seeds, with and without husks in a petri-dish would indicate whether it is preferable to peel the husks before planting and whether or not to germinate them in the laboratory and transfer them later to the field.

Another point is how to maintain the genetic diversity of the wild accessions. As already pointed out, it is recommended to collect a single dispersal unit from an individual wild oat plant and from several dozens of plants from any particular site. Maintaining the genetic diversity in the seed sample requires that at the time of regeneration, all the seeds, and not just a sample of them, should be planted. Again, seeds of these plants must be harvested separately and the new seed yield also should be kept that way. The result is that each accession of wild oat may grow to several dozens of sub-accessions.

3.3.4.1 Characterization and Utilization

Characterization of Gene Bank accessions usually refers to the description of morphological and physiological characters that can be recorded at multiplication and regeneration phases. Looking into characteristics that have utilization value may be regarded as evaluation. There is some disagreement between Gene Banks and breeder regarding the responsibility for the evaluation which usually requires special skill, facilities, and funding.

Utilization is the incorporation of specific gene, or genes, of an accession in breeding lines and newly bred varieties. When it comes to utilizing genetic resources, breeder always would prefer cultivated germplasm, wild species is the last resort and if required those of the crop primary genepool are preferred. The reason is that gene transfer via hybridization mix characters of domestic and wild nature and several back-crosses are required to select the domesticated form with the desired character of the wild species. The difficulty with species of the crop's secondary genepool is the various degrees of sterility in the F1 hybrids and even in segregating generations.

In oat, however, wild relatives have traditionally been an important source, and sometimes the only source, for genetic diversity particularly to biotic and abiotic stresses. Most probably it will remain so in the future. Many oat cultivars and breeding lines have wild relatives in their pedigree that contributed mainly genes for disease resistance. Most notably are genes for crown rust and stem rust resistance from ssp. *sterilis* and *fatua* but also from *A. barbata*. Resistance to cover smut was found in ssp. *sterilis* and against loose smut in ssp. *fatua* and *A. barbata*.

Yet, a legitimate question is how many oat cultivars have been generated by hybridization with wild relatives. While that kind of information is not readily available, it is quite safe to say that nearly all the crown rust resistant oat cultivars carry genes extracted from ssp. *sterilis* and *fatua*. Detailed information is nevertheless not easy to get. Unfortunately, Gene Banks do not keep in their database records on domesticated oat cultivars and breeding lines that possess genes from wild relatives. Because of that breeders cannot fully appreciate the usefulness of a particular domesticated accession for his program. Not less important is that this kind of information would clearly reflect the value of collecting and maintaining wild genetic resources in Gene Banks.

The domestication of *A. magna* is (see Sect. 2.5.2) may open additional avenues in oat breeding. Future development of this highly protein rich oat will require additional gene transfer from the common oat and from wild *A. magna*. The oat wild genetic resources maintained in gene banks throughout the world are indispensable for such an endeavor.

Glossary

Acuminate Sharp-pointed

Allopolyploid An organism with more than two sets of chromosomes

Amphidiploid Synonym of allopolyploid

Anaphase A stage of mitosis and meiosis

Aristate Terminates by a slender bristle

Autopolyploid An organism having more than two sets of homologous chromosomes

Autosyndesis Pairing between chromosomes of the same parent

Awn A rigid bristle issuing from the back of the lemma

Callus A thickened structure at the base of the floret

Cotyledon The primary leaf in the embryo

Cultigen Domesticated plant

Denticulate Finely dentate

Diploid An organism having double sets of chromosomes

Epiblast A rudimentary second cotyledon

Epistasis The pre-dominanting of two characters whose genes are not allomorphs

Flag leaf The leaf below the panicle

Floret Individual flower of grasses

Genome The chromosomes of the gamete

Glumes The two outer chaffy bracts of the spikelet

Hillum A scar left when the seed breaks from its stalk

G. Ladizinsky, *Studies in Oat Evolution*, SpringerBriefs in Agriculture, DOI: 10.1007/978-3-642-30547-4, © The Author(s) 2012

Homoeologus Partly homologous

Keel Projecting longitudinal ridge

Lanceolate Broadened at the base and tapering toward the apex

Lemma The lower bract of the grasses flower

Meiosis Process of reduction division in germ-cells

Metaphase The stage in mitosis or meiosis in which chromosomes split up in equatorial plate

Monosome A line lacking one chromosome

Nullisome A line lacking a pair of homologous chromosomes

Ovate Egg-shaped

Out-crossing Fertilization by gametes of other plants

Panicle The oat inflorescence

Peduncle Leafless axis bearing one or several flowers

Polyploid An organism with reduplication of chromosomes

Prophase The first stage of mitosis or meiosis

Rachilla The spikelet axis

Subulate Tapering from base to apex

Taxon (pl.taxa) A taxonomic unit

Tiller A stalk issuing from the bas e of the plant

Telophase Final stage of mitosis and meiosis

Ventricose Swollen condition

Villose Bearing moderately soft hairs

References

Aung T, Zwer P, Park P, Davies P, Sidhu P, Dundas I (2010) Hybrids of *Avena sativa* with two diploid wild oats (CIav956) and CCIav7233) resistant to crown rust. Euphytica 174:189–198

Baum BR (1977) Oats: wild and cultivated. Minister of Supply and Service, Ottawa

Baum BR, Fedak G (1985a) *Avena atlantica*, a new diploid species of the oat genus from Morocco. Can J Bot 63:1057–1060

Baum BR, Fedak G (1985b) A new tetraploid species of *Avena* discovered in Morocco. Can J Bot 63:1379–1385

Cahana H, Ladizinsky G (1978) The cytogenetic position of *A. damascene* among the diploid oats. 1978. Can J Genet Cytol 20:399–404

Chen Q, Armstrong K (1994) Genomic in situ hybridization in *Avena sativa*. Genome 37:607–612

Coffman FA (1961) Oats and oats Improvement. American Society of Agronomy, Madison, pp 650

Feldman M, Strauss I (1983) A genome restructioning gene in *Aegilops longissima*. In: Proceedings of the 6th International. Wheat Genetics Symposium. Kyoto, Japan, pp 309–314

Gandoger MM (1907) Florule de Ceuta (Maroc). Bull Soc Bot Fr 54:77–81

Gandoger MM (1908) Florule du litoral Mediterraneen. Bull Soc Bot Fr 55:656–659

Harlan JR, de Wet JMJ (1971) Toward rational classification of cultivated plants. Taxon 20:509–5517

Heneen WK (1963) Extensive chromosome breakage occurring spontaneously in certain individuals of *Elymus fractus* (= *Agropyron junceum*). Hereditas 49:1–32

Jakubziner MM (1958) New wheat species. In: Jenkins BC (ed) Proceeding 1st International wheat genetics symposium. University of Manitoba, Winnipeg, pp 207-220.

Jellen EN, Gill BS, Cox TS (1994) Genomic in situ hybridization differentiates between A/D and C genomes chromatin and detects intergenomic translocations in polyploidy oat species. Genome 37:613–618

Jones ET (1940) A comparison of the segregation of wild versus normal or cultivated base in grains of diploid, tetraploid and hexaploid oats. Genetica 22:419–434

Katsiotis A, Hgadimitriou M, Heslop-Harrison JS (1997) The close relationships between the A and B genomes in Avena L. (Poaceae) determined by molecular cytogenetic analysis of total genomic tandemly and repetitive DNA sequences. Annl Bot 79:103–109

Ladizinsky G (1968) Cytogenetic and evolutionary study in the genus *Avena* L. Ph.D thesis submitted to the Hebrew University (Hebrew with English summary)

Ladizinsky G (1971) *Avena prostrate*: A new diploid species of oat. Israel J Bot 20:297–301

G. Ladizinsky, *Studies in Oat Evolution*, SpringerBriefs in Agriculture, DOI: 10.1007/978-3-642-30547-4, © The Author(s) 2012

Ladizinsky G (1973) Genetic control of bivalent pairing in *Avena strigosa* polyploidy complex. Chromosoma 42:105–110

Ladizinsky G (1974) Genome relationships in the diploid oats. Chromosoma 47:109–117

Ladizinsky G (1975a) Collection of wild cereals in the upper Jordan Valley. Econ Bot 29:264–267

Ladizinsky G (1975b) Oats in Ethiopia. Econ Bot 29:238–241

Ladizinsky G (1975c) On the origin of broad bean Vicia faba L. Israel J Bot 24:80–88

Ladizinsky G (1993) The taxonomic position of *Avena magna*—reappraisal. Lagascallia 17:325–328

Ladizinsky G (1995) Domestication via hybridization of the wild tetraploid oats *Avena magna* and *A. murphyi*. Theor Appl Genet 91:639–646

Ladizinsky G (1998) A new species of *Avena* from Sicily, possibly the tetraploid progenitor of hexaploid oats. Genet Res Crop Evol 435:263–269

Ladizinsky G (1999) Identification of the lentil's wild genetic stock. Genet Res Crop Evol 46:115–118

Ladizinsky G, Zohary D (1968) Genetic relationships between diploids and tetraploids in Series Eubarbatae of *Avena*. Can J Genet Cytol 10:68–81

Ladizinsky G, Zohary D (1971) Notes on species delimitation, species relationships and polyploidy in *Avena* L. Euphytica 20:380–395

Ladizinsky G, Jellen EN (2003) Cytogenetic affinities between populations of *Avena insularis* Ladizinsky from Sicily and Tunisia. Genet Res Crop Evol 50:11–15

Leggett JM (1998) Chromosome and genomic relations between the diploid species *Avena strigosa, A. eriantha* and the tetraploid *A. maroccana*. Heredity 80:361–367

Leggett JM, Ladizinsky G, Hagberg P, Obanni M (1992) Distribution of nine *Avena* species in Spain and Morocco. Can J Bot 70:240–244

Leggett JM, Thomas HM, Meredith MR, Hamphry MW, Morgan WG, Thomas H, King IP (1994) Intergenomic translocations and the genome composition of Avena maroccana Gdgr. revealed by FISH. Chromosom Res 2:163–164

Leggett JM, Markhand GS (1997) A revision of genome evolution in hexaploid Avena ? The Aberystwyth Cell Genetic Group, 7th Annul Meeting. Poster no. 21

Lilienfeld FA, Kihara H (1951) Genome analysis in *Triticum* and *Aegilops* X. Concluding review. Cytologia 16:101–123

Maire R (1953) Flore de l'Afric du Nord. Paul Lechevalier (ed)

Malzew AI (1930) Wild and cultivated oats section *Euavena Griseb*. Bull Appl Bot Genet Plant Breeding (USSR) Suppl. 38

Martinoli G (1955) Cytotaxonomy of some species of the genus *Avena* from Sardinia. Caryologia 7:191–204

Martinoli G (1969) *Avena*: new oat species. Science 163:594–595

Morris R, Sears E R (1967) The cytogenetics of wheat and its relatives. In: Quisenberry KS, Reitz PL (eds) Wheat and Wheat Improvement. American Society of Agronomy Madison. PP 19–87

Murphy HC, Sadanaga k, Zillinsky FJ, Terrell EE, Smith RT (1968) Avena magna: an important new tetraploid species of oats. Science 159:103–104

Nishiyama I (1929) The genetics and cytology of certain cereals. Morphological and cytological studies on triploid, pentaploid and hexaploid *Avena* hybrids. Jpn J Bot 5:1–48

Nishiyama I (1936) Cytogenetical studies in *Avena* I. chromosome associations in hybrids between *Avena barbata* Pott and autotetraploid of *A. strigosa* Schreb. Cytologia 7:276–281

Rajhathy T (1959) A reexamination of the chromosomes of the tetraploid '*Avena sterilis*' from Sardinia. Can J Genet Cytol 11:1001–1004

Rajhathy T, Thomas H (1967) Chromosomal differentiation and speciation in diploid *Avena*, III, Mediterranean wild populations. Can J Genet Cytol 9:52–68

Rajhathy T, Baum BR (1972) *Avena damascena*: a new diploid oat species. Can J Genet Cytol 14:645–654

Rajhathy T, Morrison JW (1959) Chromosome morphology in the genus *Avena*. Can J Bot 37:331–337

Rajhathy T, Morrison JW (1960) Genome homology in the genus *Avena*. Can J Genet Cytol 2:278–289

Rajhathy T, Sadasivaiah RS (1968) The chromosomes of *Avena magna*. Can J Genet Cytol 10:385–389

Riley R, Chapman V (1958) Genetic control for the cytologically diploid behavior of hexaploid wheat. Nature 182:713–715

Sears ER (1954) The aneuploids of common wheat. Missouri Agric Stn Res Bull 572:1–58

Thomas H (1992) Cytogenetics of avena. In: Marshall G, Sorrell ME (eds) Oat Science and Technology. American Society of Agronomy, Madison

Zohary M, Feinbrun N (1953) Analytical flora of the land of Israel (in Hebrew)